W9-BPM-104

ROOFING
INSTANT ANSWERS

DATE

ROOFING INSTANT ANSWERS

Terry Kennedy

McGRAW-HILL

New York Chicago San Francisco Lisbon London
Madrid Mexico City Milan New Delhi San Juan
Seoul Singapore Sydney Toronto

McGraw-Hill

*A Division of The **McGraw·Hill** Companies*

1 2 3 4 5 6 7 8 9 0 DOC/DOC 0 9 8 7 6 5 4 3 2

ISBN 0-07-138712-9

The sponsoring editor for this book was Larry S. Hager and the production supervisor was Sherri Souffrance. It was set in Stone Sans by Lone Wolf Enterprises, Ltd.

Printed and bound by R. R. Donnelley & Sons.

This book is printed on recycled, acid-free paper containing a minimum of 50% recycled, de-inked fiber.

CONTENTS

ACKNOWLEDGMENTS

This book is dedicated to all of the people over many centuries who have been out there on the roof deck, doing the hard work. Roofing is like oxygen—society takes it for granted. But in the end, the most commonplace things serve the most important functions in our lives. Without roofs, our lives would simply not work. There is very little thanks given to the countless people who have been out there on the deck, keeping us dry and cozy; and to all of them let's offer up a round of applause.

I would also like to thank Larry Hager for all of his help and Shelly Mitchell for being such a great friend and her always giving fearless, and continual inspiration and support.

Terry Kennedy

ABOUT THE AUTHOR

Terry Kennedy has been an AEC (architecture, engineering, contracting) professional for over 35 years. He has installed many roofs working on the deck, and been responsible for the installation of many others. In addition to his field experience, he has consulted in countless settlements for lawsuits that evolve from roofing defects. This insight gives the author a strong hand in recognizing what is really needed when it comes to being true roofing professional.

BEFORE THE JOB STARTS

T he roofing contractor or the estimator should use this section of *Roofing Instant Answers* to expedite the tasks that arise before mobilization: the site visit, take-offs, analysis of what the job will entail, estimates, schedules and contracts.

WHEN A REQUEST TO BID COMES IN

- A new job should be considered to have started when a request to bid is received.

- The contractor should be prompt and efficient, but should not hurry this important phase of a job.

- The well-run office has a thorough system for handling estimating, planning and contracts.

- All requests to bid should be delivered to the same workstation for punching into the company's computer system. This helps to avoid data entry errors.

- The best person to start the job trail in the computer system is whoever is in charge of entering new jobs in accounting.

- Treat a request to bid as if a new job has started.

- The data is entered into the job cost accounting software as a pending account.

- The person in accounting enters the job and posts payables, such as estimating expenses, against the new account.
- Enter the data once—keystrokes cost money.
- Have your computer consultant integrate software for accounting, estimating, scheduling, etc., so that after the request to bid is entered into the system, the data from the account is automatically entered when estimating, scheduling and mobilization begin.
- Keep a running update on what your estimates have been in relationship to the actual profits made on your jobs.
- Use the different areas of the estimate in which you have lost money as a guide for developing a list of what you need to be especially careful of during estimating.

The First Site Visit

- After the first site visit, the information gathered from looking at the location will become an important part of take-offs from plans, estimating, pre-contract discussions, developing the contract document and pre-planning.
- The estimator should imagine applying the roof during the site visit.
- Check the condition of the roof and make notes.
- Make all measurements while on the roof.
- Search carefully for any prep work that needs to be done.

fastfacts

➤ *The only inappropriate question when you are preparing to bid is the one that you do not ask.*

➤ *Learning to pass on jobs is hard to do, but it is as important as getting a job.*

➤ *In order to be fully effective in making a job timely and profitable, the estimator or roofing contractor must begin the project with a visit to the site.*

- Check the condition of vents, vent caps and flashings.
- Check the condition of the chimney, if there is one.
- Check for potential water leakage into the house.
- Notice anything that might cause delays or tie up your work.
- Check for unexpected labor costs.
- Determine if the rafters are secure.
- Determine the number of roofs already on the structure.
- Determine the appropriate shingle or membrane if you are adding a roof.
- Is a tear-off necessary for an accurate bid?
- Once you are back on the ground, discuss your findings with the homeowner or project manager.
- Look carefully and think about any potential hidden repair work that may present unforeseen expenses.
- Ask the owner about any necessary prep-work, such as protecting belongings or inhabitants, any proposed supplementary work, and any hidden repair work.

Labor and material take-offs will be the prime concern at this time, but analysis of how logistics at the site will affect the project must also take place.

The checklist in Figure 1.1 is for the roofing contractor or the estimator to take to the site on the first visit. This checklist can be adjusted to the specific needs of your company. Then it can be used for all types of jobs and dispatched with the estimator.

Back In The Office

- The entire roofing business should be interlinked via the computer system.
- As jobs come in, they are entered into the database as a temporary account.
- They are estimated.
- If they are won, they become a new account.
- If the job is lost at the bid stage, the information about the client remains in the marketing database.
- Announcements about your firm are then sent to the principals of that company on a regular basis.

CHECKLIST FOR THE FIRST SITE VISIT		
✔	ITEM	NOTES
	THE BUILDING	
	How many stories is the structure?	
	What type of roof is to be installed?	
	Is the carpentry sound and complete?	
	What types of wall sheathing are on the building?	
	What is the general design of the roof?	
	Are there complex intersections on the roof?	
	What type of sheathing is on the deck?	
	EASE OF WORK	
	Is the access simple?	
	Is there a secured stockpile area for materials?	
	Is there room for materials lifting?	
	Are utilities readily accessible?	
	Is the area above the roof deck open?	
	Are there visible hindrances to performance?	
	Does the general contractor appear to be well organized?	
	If not, what types of impedance of the roof installation should be anticipated?	
	If the take-offs are from plans, what is known about the general contractor?	

FIGURE 1.1   Sample checklist for first site visit.

- If the preliminary bid is accepted and set up as a new account, a tentative schedule has already been roughed out.
- Every step of the process is entered in computer, and they should all relate to one another.

Take-Offs

- The building should be built in the estimator's imagination during take-offs.
- Even if you are bidding straight off a set of plans, you must visit the site and review the logistics that the mechanics will run into at the location.
- Spend the money for a digitizer that will measure the areas on plans and dump them straight into your computer.
- Always review take-offs.
- Do all measurements and extend all counts at least twice.

ESTIMATING GUIDELINES

Estimating is a very important part of the roofing industry. If the contractor and the estimating team are not fast, a lot of money can be thrown at estimating jobs that never go into contract.

On the other hand, if the contractor and the estimating team are not accurate, a company can go belly up from just a handful of jobs.

To Estimate Labor Costs Properly, Consider Two Components

- ✔ Work hours required to perform the installation
- ✔ Taxes and insurance costs associated with the total work hours allocated to the job

- Estimate each segment of the installation separately.
- Include any and all work required to correct imperfections.
- Include work needed to ensure that the structure, substrates, and projections meet code requirements; this may account for as many work hours as the installation of the roofing materials.

- To calculate labor costs, multiply the labor rate per hour by the total number of work hours.
- Once the total number of work hours is calculated, multiply that figure by the labor rate per hour to estimate the labor costs for completing the basic installation for each segment of the job.
- Combine this formula and the materials estimate to arrive at the costs for the individual job segments.
- Sealing around projections, forming and installing flashings and valleys, and installing other finishing material all fall into the category of finishing labor.
- Roofing contractors' familiarity with their crews, along with accurate records of the amount of time each crew requires to perform tasks, enables them to properly estimate the number of work hours needed for each step.
- As mentioned earlier, labor amounts are actually composed of two sets of costs.
- The labor burden must be added to the work hours to calculate total labor costs.
- On average, the labor burden increases each job's labor costs by 25 to 30 percent.
- The labor burden is the percentage of payroll dollars roofing contractors are compelled to pay to government agencies in the form of taxes, and to insurance companies in the form of premiums.

The Estimating Checklist

Figure 1.2 is a handy reference for estimators to make certain that estimates are complete before they are submitted. Be sure to include environmental factors in your estimates. They become more important every year, and can make or break your profit margins.

Environmental Factors To Consider When Bidding A Job

✔ The cost of disposing the waste in a local landfill
✔ Will debris from the job need to be sorted for disposal?
✔ If sorting is required, how many work hours are required?
✔ Can all debris be disposed of in normal fashion?
✔ If not, how is disposal accomplished? At what cost?

ESTIMATING CHECKLIST		
✔	ITEM	NOTES
	THE BUILDING	
	How many stories is the structure?	
	What type of roof is to be installed?	
	Is the carpentry sound and complete?	
	Does the general contractor appear to be well organized?	
	If not, what types of impedance are noticeable?	
	If the take-offs are from plans, what is known about the general contractor?	
	Is the street use clear?	
	Are there government requirements?	
	What is the ability of the crew that will be applying the roof?	
	What is the local weather?	
	Quality of the local labor pool?	
	What is the terrain at the site?	
	Nature of the built environment around the structure?	
	Unique access requirements?	
	Recycling requirements?	
	Dump requirements?	
	Toxic waste disposal needed for job?	
	EASE OF WORK	
	Is the access simple?	
	Is there a secured stockpile area for materials?	
	Is there room for materials lifting?	
	Are utilities readily accessible?	
	Is the area above the roof deck open?	
	Are there visible hindrances to performance?	

FIGURE 1.2   Estimating checklist.

CALCULATING ROOF AREAS

A Simple Shed Roof

- Most roofs are comprised of one or more rectangles (Figure 1.3).
- The area of the entire roof is the sum of the areas of each rectangle.
- For example, the area of a shed roof, which has only one rectangle, is found by multiplying the rake line by the eave line, or A x B.

Gable Roof

- The gable roof is comprised of two rectangular planes.
- Its area is found by multiplying the sum of the rake lines by the eave lines, or A x (B+C).

Gambrel Roof

- Gambrel roofs have four rake lines.
- The total area is calculated by multiplying the sum of the rake lines by the eaves lines, or A x (B+C+D+E).

Complicated Roofs

No matter how complicated a roof might be, projecting it onto a horizontal plane easily defines the total horizontal surface.

- Figure 1.4 illustrates a typical roof complicated by valleys, dormers, and ridges at different elevations.

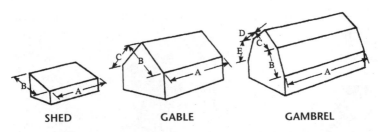

SHED   GABLE   GAMBREL

FIGURE 1.3 Simple roofs.

- The lower half of the figure shows the horizontal projection of the roof.

- In the projection, inclined surfaces appear flat and intersecting surfaces appear as lines.

- Measure the roof's horizontal projection from the building plans, from the ground, or from inside the attic.

- Once the measurements are made, draw the horizontal area of the roof to scale and calculate the area.

- Because the actual area is affected by the slope, calculations must be grouped in terms of roof slope.

- Measurements of different slopes are not combined until the true roof areas have been determined.

- The horizontal area under the 9 inch slope roof is:

  $26 \times 30 = 780 + 19 \times 30 = 570$

- Total = 1350 square feet

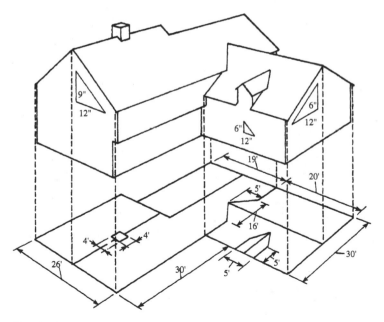

FIGURE 1.4   Horizontal projection of a complex roof.

- From this gross figure, deductions can be made for the chimney area and for the triangular area of the ell roof that overlaps and is sloped differently from the main roof.
- The chimney area = 4 x 4 = 16
- Joining roof ½(16 x 5) = 40 (triangular area where room addition joins the main house).
- Total = 56 sq. ft.
- Then the net projected area of the main roof can be calculated.
- 1350 − 56 = 1294 sq. ft.
- The calculation for the horizontal area under the 6 inch slope roof is:

  20 x 30 = 600

  ½(16 x 5) = 40

  Total = 640 sq. ft.
- In the example, there are three overlapping areas:
  - On the 6 inch slope roof where the dormer eaves overhang 2 feet.
  - On the 9 inch slope roof where the main roof eaves overhang 3 feet.
  - The areas where the main roof eaves overhang the smaller section of the main roof in the rear of the building.
- In each case, if the eaves extend 4 inches beyond the structure, the duplication calculations are:
  - Two eaves overhangs, 6 inch slope:

    5 x 4/12 = 20/12 x 2 = 3⅓ sq. ft.

fastfacts

➤ *Portions of the higher roof planes often project over roof surfaces below them, but the horizontal projections do not show the overlap.*

➤ *Add these duplicated areas to the total horizontal area.*

➤ *Make one final correction to account for these overlapped or duplicated areas before the total projected horizontal area is obtained.*

- Two eave overhangs, 9 inch slope:
- 7 x 4/12 = 28/12 x 2 = 4⅔ sq. ft.
- Overhang covers only half of the 19-foot wide section:
- 9.5 x 4/12 = 3⅙ sq. ft.
- Item one is added to the area of the 6 inch slope roof.
- Items two and three are added to the 9 inch slope roof.
- Thus, for the 6 inch slope roof, the adjusted total is
  640 + 3 = 643 sq. ft.
  and for the 9 inch slope roof:
  1294 + 8 = 1302 sq. ft.

When the total projected horizontal areas for each roof slope are calculated, convert the results to actual areas with the aid of Table 1.1. To use the table, multiply the projected horizontal area by the conversion factor for the appropriate roof slope. The result is the actual area of the roof. For example, the ell roof overhang for the 9 inch slope roof:

- The Horizontal x Conversion = Actual
  1302 sq. ft. x 1.250 = 1627.5 sq. ft.

For the 6 inch slope roof:

- The Horizontal x Conversion = Actual
  643 sq. ft. x 1. 118 = 718.8 sq. ft.

After you convert the horizontal areas to actual areas, add the results to obtain the total area of roof to be covered:

1628 + 719 = 2347 sq. ft.

In this case, assume a 10 percent waste allowance.

The total area of roofing material required is:

2347 + 235 = 2582 sq. ft.

Remembering Additional Materials

- To complete the estimate, determine the required quantity of starter strips, drip edges, hip and ridge shingles, and valley strips needed for the job.

fastfacts

➤ *Fractions are rounded off to the nearest foot.*

➤ *In the actual job estimate, include an allowance for waste.*

➤ *The same horizontal area projection and roof slope always results in the same actual area regardless of roof style. In other words, if a shed, gable or hip roof, with or without dormers, covered the same horizontal area and had the same slope, they would each require the same area of roofing to cover them.*

- Each of these quantities depends on the length or size of the eaves, rakes, hips, ridges, and valleys.
- Since eaves and ridges are horizontal, their lengths can be measured from the horizontal projection drawing.
- Rakes, hips and valleys are sloped.
- Their lengths must be calculated following a procedure similar to that for calculating sloped roof areas.

Rakes

- To determine the true length of a rake, first measure its projected horizontal distance.
- Table 1.1, used to convert horizontal areas to actual areas, can be used to convert rake lengths.
- To use the table, multiply the rake's projected horizontal length by the conversion factor for the appropriate roof slope.
- The result is the actual length of the rake.
- For the house shown in Figure 1.5, the rakes at the ends of the main roof have horizontal distances of 26 feet and 19 feet.
- There is another rake in the middle of the main house where the higher roof section meets the lower.
- Its horizontal distance is: 13 + 3.5 = 16.5 ft.
- Adding all these horizontal distances gives a total of 61.5 feet.

TABLE 1.1 Estimating Areas Using Roof Pitches and Inclined Areas

Classification	Incline	
	Inch per foot horizontal	Angle with horizontal
Flat roofs	⅛	0°-36'
	¼	1°-12'
	⅜	1°-47'
	½	2°-23'
	⅝	2°-59'
	¾	3°-35'
	1	4°-45'
	1⅛	5°-21'
	1¼	5°-57'
	1½	7°- 8'
	1¾	8°-18'
	2	9°-28'
Steep roofs	2¼	10°-37'
	2½	11°-46'
	2¾	12°-54'
	3	14°- 2'
	3¼	15°- 9'
	3½	16°-16'
	3¾	17°-21'
	4	18°-26'
	4¼	19°-30'
	4½	20°-34'
	5	22°-37'
	6	26°-34'
	7	30°-16'
	8	33°-42'
	9	36°-52'
	10	39°-48'
	11	42°-31'
	12	45°- 0'
Extra steep roofs	14	49°-24'
	16	53°- 8'
	18	56°-18'
	20	59°- 2'
	22	61°-23'
	24	63°-26'

- Using Table 1.1, the actual lengths for the 9 inch slope roof are calculated as follows:
- Horizontal x Conversion = Actual

  61.5 sq. ft. x 1.250 = 76.9 sq. ft.
- Following the same procedure for the ell section, with its 6 inch slope roof and dormer, the total length of the rakes is found to be 39 feet 1 inch.
- Add these rake lengths to the total length of eaves, which is the actual horizontal distance.
- No conversion is necessary.
- This provides the quantity of drip edge required for the job.
- The quantity of ridge shingles required is estimated directly from the drawing since ridgelines are true horizontal distances.

Hips and Valleys

- Hips and valleys also involve sloped distances.
- Convert their projected horizontal lengths to actual lengths.
- Measure the length of the hip or valley on the horizontal projection drawing and multiply it by the conversion factor to get the appropriate roof slope.
- The result is the actual length of the hip or valley.
- On the illustrated house in Figure 1.5, there is a valley on both sides of the ell roof where it intersects with the main roof.
- The total measured distance of these valleys on the horizontal projection is 16 feet.
- The fact that two different slopes are involved complicates the procedure.
- If there were only one roof slope, the true length could be calculated.
- Since this house has two slopes, calculations for each slope must be made and then averaged to obtain a close approximation of the true length of the valleys.

Horizontal x Conversion = Actual

16 ft. x 1,600 = 25.6 ft. for the 9 inch slope roof

16 ft. x 1,500 = 24.0 ft. for the 6 inch slope roof

- The approximate length of the two valleys is 24.8 feet or 12.4 feet each.
- The total projected horizontal length of the dormer valleys in Figure 4.2 is 5 feet.
- With the ell roof and the dormer both having 6 inch slopes, the actual length of the valleys is calculated to be 7.5 feet.
- The total length of the valleys for the house is: 25 + 7 = 32 feet.

WRITING CONTRACTS

- Review all standard form contracts at regular intervals to make sure that the language keeps pace with legal requirements and industry changes.
- Any matter that is unsettled by the contract is a potential lawsuit.
- An accurately drafted roofing contract can reduce or eliminate litigation by providing, in advance, solutions to problems that might arise during the course of the roofing job.
- Along with supplying solutions to problems, the contract should put in writing the understandings of all parties concerned with the roofing job.
- Stipulate all items very carefully; this is a period when taking your time can make a big difference.
- A preliminary draft of the document should have been presented to the client before the pre-roofing conference.
- When the preliminary draft hits the conference table, all parties can voice their conflicting viewpoints and iron out acceptable solutions.

PRE-PLANNING

- Planning is the most important phase of a roofing job after a bid is accepted.
- Planning always begins while the estimate is being calculated.
- During estimating, the estimator and the project manager begin to plan the project.

- The more care put into planning, the more profits and word-of-mouth marketing a company will enjoy.
- Do not confuse planning with scheduling.
- Planning deals with the strategy of how the tasks involved with the job will happen.
- Planning continues through scheduling, pre-contract meetings, and preparing contracts.
- Planning begins when the request to bid is first entered in the system.

Pre-Roofing Conferences

- Pre-roofing conferences are an open discussion between all parties that have anything to do with the roof.
- Bring all parties into the discussions: the owner, architect, engineers, roofing contractor, deck contractor, siding contractor, framer, general contractor, roofing materials manufacturer, insulation manufacturer, and representatives from any trades that have a reason to be on the roof.
- The purpose of these meetings is to enable all parties to iron out as many of the logistics required for working together on the project as possible.
- The meetings offer everyone a chance to strive for the best deal for their own firm and clarify many items that would be difficult to come to terms over during the heat of construction.

fastfacts

➤ *The dynamics of today's roofing industry make pre-roofing conferences more important than ever before.*

➤ *If differences cannot be worked out, or you will not be properly compensated for differences in opinion about the job conditions, it is time to walk away.*

fastfacts

Establish a company policy that will clarify potential conflicts during pre-contract negotiations. This can be a big step toward continual profits for your business and for preventing construction defect litigation.

Delegating Responsibility

- Important aspects of pre-construction meetings is delegating authority for each task, and hammering through what is to happen if the schedule backs up.
- Put the agreements achieved at this meeting on paper.
- Documents in writing can be a very strong tool for three major problems that can arise after a project hits the roof deck:
  - Disputes over a lack of clarity of details, and assigned responsibility for tasks at complex areas of the roof.
  - Disputes over construction management and the overlapping of the trades, and breakdowns in communication.
  - Cost overruns.
- Pre-construction meetings contribute to clarifying the details of the roofing contract, which stipulates procedures and protections for the roofing firm.
- All parties involved in the design and construction of the roof system must recognize and meet their responsibilities to achieve a proper roofing system.
- No link must be broken in this chain of responsibility.

Owner Responsibilities

- The owner is the first link in the responsibility chain.
- The owner's first obligation is to achieve a proper roofing system for the building.
- The best roof, as determined by the architect, must be accepted and paid for by the owner.

- Most owners are not knowledgeable about roofing systems and specifications.
- Usually they defer to the architect's recommendations.
- The owner has the additional responsibility of not limiting the architect in the selection of the roofing system and insulation purely because of budgetary concerns.
- Decisions based solely on price often result in failed roofs.
- Once a roof system is accepted, the owner's responsibilities become more limited until the building is complete and the maintenance program begins.
- The owner's maintenance responsibilities can't be overemphasized.
- An ongoing maintenance program must be developed and followed throughout the life of the building.
- Ignoring roof maintenance is the same as absolving all other parties of their responsibilities.

Architect Responsibilities

- Legal authorities claim that roofs and foundations are the two structural components most likely to cause a lawsuit involving the architect.
- The architect's motive for designing the proper roofing system is twofold: self-preservation and concern for the client's interests.
- Get to know the architects who do business in your market area.
- Speak to other contractors to determine the architects who consistently produce quality work with accurate details, and those who produce designs that are refused by other contracting firms.
- Challenge the architect's plan if it does not meet with generally accepted roofing practices.
- If the plans are sound, the architect will have no trouble clarifying any questions about the design elements of the structure.

Manufacturer Responsibilities

- All roofing materials have limitations.
- Manufacturers know the limitations of the product and are required to represent their line accurately.
- Manufacturers must be willing to certify their materials.

- If they meet the appropriate standards, certification is reasonable.
- If the materials do not have standards, that should be verified by the manufacturer.
- With materials such as bitumens, certification to standards is a must.

Roofing Contractor Responsibilities

- The idealized concept of the perfect roofing system has never occurred.
- With any construction project there are so many factors involved, that situations may arise, which some might consider flaws.
- Roofing is not performed in a hermetically sealed laboratory. It is vulnerable to dozens of rugged conditions.
- The responsibility of the roofing contractors is to recognize the many difficult situations specific to the structure, and then to make every attempt to overcome those problems.
- The basis of any high-quality roofing firm is knowledgeable, conscientious supervisory personnel and properly trained work crews.
- Organized apprenticeship programs can result in work crews that are trained in the contractor's procedures to ensure consistent performance on every job, regardless of which crew performs the work.
- Contractors must inform the architect, in writing, of any recommended changes that they feel are in the best interest of the project.
- This notification should be made immediately upon discovering that a change is required.
- Never install a roof that is inadequate; your integrity is always a very important part of your business life.
- Nothing causes business to drop off faster than a failed roofing system and an unhappy client.

What Should Be Accomplished at Pre-Roofing Conferences

- At the pre-roofing conference, a definite game plan should be pursued.
- This is a chance to ask all the tough questions that make or break the profit line after the job begins.

- Avoiding the opportunity to hammer out the vital questions before hitting the roof deck is probably the gravest error any roofer can make.

- It is important to remember that you should approach the pre-roofing meeting positively, rather than competitively.

- Use it as a welcome opportunity to come to an understanding and to sell services that raise the gross value of the project.

- In the last few decades, owners, architects, and designers have become much more aware of the importance of roofs.

- For this reason, as well as to eliminate callbacks and repair jobs caused by owner neglect, the pre-roofing conference is the perfect time to establish a maintenance program with the owners or their representatives.

- When a pre-roofing conference is utilized to the maximum, the benefits are widespread:
    - All parties get what they want.
    - The owner and architect are satisfied that they are getting the best possible roof.
    - The manufacturer is satisfied that the installation meets warranty requirements.
    - The general contractor, construction manager, and the roofing contractor get the roof installed with few problems, and all concerns are coordinated with the other trades.

VALUE ENGINEERING

- Value engineering sounds like some kind of a complicated system that might come from a structural analysis of the building, but typically the phrase is a marketing and sales expression.

- In reality it refers to cost cutting ideas,that the general contractor and subs can offer to the principal and design team.

- It should be remembered that the point is to find high quality methods of attaining cost savings, not questionable methods that may lead to an inferior product. This is very important.

- Replacing the existing design development concepts with inferior quality work and material can come back in the form of legal ramifications.

CHECKLIST FOR PRE-ROOFING CONFERENCES		
✔	ITEM	NOTES
	All questions about the drawings, details and specifications have been listed for discussion.	
	All questions, restrictive clauses and contingencies related to the condition of the roof have been listed.	
	All prep work that needs to be done has been listed.	
	All vent, vent cap, chimneys and flashing conditions have been listed for the meeting.	
	All questions related to specified products have been listed.	
	All past and potential water leakage into the structures has been listed.	
	All potential delays, including access, weather, materials acquisition, overlapping of trades, site access, framing repair, etc., which may tie up your work have been addressed.	
	Questions related to unexpected labor costs have been listed.	
	All cleanup issues, including environmental and recycling, have been addressed.	
	All opportunities to sell extras and change orders have been addressed.	
	Value engineering list is ready for meeting.	

FIGURE 1.5 Checklist for pre-roofing conferences.

- Here is a common example of a value engineering offer, which might very well overcome the competition and land a contract for your firm:
  - You discover that your suppliers have a line of product that is being discontinued, but is of equal quality to that which is specified.
  - The vendor assures you that there is plenty of extra product for repairs, which may transpire before the life of the roof has cycled.
  - There is a place for the owner to stockpile replacement material and the savings is considerable.
  - You inform the general contractor (never go around the GC to the building owner or architect; your integrity is very important to long-term business and your own personal sense of well being).
  - The owner save money and becomes a satisfied customer.

Items to be Addressed in the Roofing Contract

- Exactly what work is to be done for the stated price?
- What, if any, clean-up services and environmental issues are included in the contract price?
- A statement that any changes to the contract must be made in writing, dated, and signed by all parties.
- A statement that the roofing contractor is not responsible for delays in completing the job due to circumstances beyond the roofing contractor's control, including, but not limited to: diverse weather, labor strikes, accidents, acts of God, site restrictions, etc.
- Terms and conditions under which payment is to be received from the owner or general contractor.
- Proof of insurance carried by the owner or general contractor.
- A clear understanding of all coverage and liability issues in relation to insurance.
- Provision for furnishing adequate job specifications to the roofer from the owner, general contractor, or their agents.
- The provision of various utilities and accessibility to the job site by the owner and/or the general contractor.
- Any penalties or interest to be assessed as a result of late payment.

- A complete understanding of how construction defects and consequential damages will be handled.

- Within the entire agreement, the roofing contractor might find language that relates to issues of risk, liability, or warranty that changes the general or supplementary conditions. The end result can be the imposition of substantial liability on the roofer and all liabilities must be clarified.

- A simple one-sentence clause that stipulates that the roofer is to verify the accuracy of the architect's plans and designs, which shifts responsibility for designs from the architect to the roofing contractor.

- The common practice of requiring the contractor to provide the manufacturer's guarantee for a specified period of time must be clarified before going into contract.

- Usually the language for manufacturer clauses is broad and does not include the same limitations and exclusions contained in the manufacturer's guarantee.

- Conceivably, this can result in the contractor being held responsible for conditions from which the manufacturer has been exempted.

- The following recommendations can help you make an informed business decision about whether or not to go forward with a particular project, and to what extent such factors should be reflected in the final bid price:

  - Examine the documents carefully and note any unusual items that impose a greater burden than would otherwise be expected by the roofing contractor.

  - Have an experienced construction attorney review the documents. The attorney should already be familiar with construction documents, so he or she should require less time to examine the documents thoroughly than a general practice attorney.

  - Assist the construction attorney by highlighting the unusual items noted above.

  - If, after consulting with the construction attorney, there are any areas of significant concern, be sure to raise these issues at the pre-roofing conference.

  - These changes and clarifications are important if it ultimately becomes necessary to defend your position in court.

fastfacts

➤ *Before you bid on a job, it is imperative that you know the risks you are assuming.*

➤ *Always remember that constructing a building is a team effort. The more that you involve all of the other players on your team—the contractor, roofer, architect, sheet metal and siding companies—the better the job will go.*

➤ *It is very important to remember this from the time the request to bid comes in, and to provide a very thorough list of questions to iron out at all pre-construction meetings.*

Scheduling

- Another benefit to pre-roofing conferences is schedule coordination between trades.

- It is not efficient to move a trade into a construction area unless the workers that are already building can keep going with their jobs.

- If they are held up waiting for another trade to finish, both time and money are lost.

- As all roofers know, if another trade, such as the framers, omit part of their work, for example, backing on a vertical wall, work is slowed and money is lost because of the down time.

- Scheduling of roofing can be done efficiently with software.

- For large firms with many big, on-going projects, such as subdivisions or massive condo clusters, a high-end package of Critical Path software may be needed.

- Remember, it is simple enough to punch out a beginning timeline of what needs to happen, but what gets complicated is re-scheduling.

- The software needs to be able to allow the scheduler to punch in changes as they come about and automatically adjust the entire schedule according these changes.

- Remember to use pre-roofing conferences to establish the time each trade needs on the roof.

fastfacts

➤ *The first schedule is simply a guideline.*

➤ *The most important part of scheduling is rescheduling.*

➤ *For a closer look at the complicated uses of critical path method, see McGraw-Hill's book,* CPM in Construction Management *by James J. O'Brien and Fredrick L. Plotnick (ISBN number 0-07-13440-3).*

- Naturally, any phase of the construction that requires co-operation of the trades should be discussed in detail, prior to the actual roofing.

- This includes the placement of mechanicals on the roof, such as air vents, skylights, plumbing vents, ductwork, and electrical intrusions.

2

JOB MOBILIZATION

Though sometimes overlooked, mobilization is an important part of any roofing job. There is often a tendency among roofing firms to rush through everything, from the estimate through demobilization. Naturally, with this mindset, mobilization will not get much attention. Use this chapter to review your mobilization policies, to make new ones, and put *Roofing Instant Answers* in all of your trucks so it can be used to launch all of your jobs.

OVERVIEW OF MOBILIZATION

- Avoid the tendency of some roofers to throw a load of roofing materials into a truck, roar over to the site, and push the crew very hard all day, trying to slap on as much product as possible.

- Do not hurry into a job and push frantically.

- Use job mobilization to review pre-planning and move onto the site correctly.

- Make sure that there have not been any major oversights during pre-planning and estimating.

fastfacts

➤ *The frantic approach breeds discontent and sloppy roofing crews.*

➤ *When oversights from estimating and pre-planning are discovered during any stage of a job, address them immediately.*

➤ *If the mistakes are minor, it is often best to bite the bullet and absorb the costs associated with fixing the problem.*

➤ *This can be a very strong marketing tool. General contractors and building owners do not like to use subs that moan about their hardships. Everyone prefers a company with a can-do attitude.*

➤ *However, if there is an extreme cost oversight during estimating and pre-planning, figure out a solution immediately, and approach the owner or the general contractor.*

➤ *Put all changes in writing and make certain that all responsible parties sign the document.*

- Review the approach devised during pre-planning to set the site up correctly.
- Reviewing at the start-up of mobilization will make the work smoother, simpler and, in the end, a carefully mobilized project will move along more quickly than a slap-dash job.
- Careful preparation at mobilization will also promote a can-do attitude among crews.

PRE-PLANNING REVIEW

In Chapter 1, "Before the Job Starts," we emphasized that planning is the most important phase of a roofing job after a bid is accepted. Roofing is a rough job, and the majority of roofing mechanics take great pride in their ability to install a quality job under all of the adverse conditions that they face.

- Management should always take the same kind of pride in their efforts to make a job run smoothly and effectively. Planning paves the way for efficient production at the project.

- The project manager should work with the estimator from the beginning. They should think through the project.

- The schedule should also be reviewed at this time.

- The pre-planning is now used to strategize how, specifically, the tasks involved with the job will take place.

- Any major oversights found in the original estimate and plan require new meetings as soon as possible.

- A meeting between all of the parties involved in the project during mobilization can be a very valuable practice.

- What is imagined during estimating, and what is found after moving onto the site, can sometimes throw the roofer some big surprises.

- Have all of the parties attend the site meeting: the owner, architect, engineers, roofing contractor, deck contractor, siding sub, framer, general contractor, roofing materials manufacturer, insulation manufacturer, and representatives from any trades that have a reason to be on the roof.

fastfacts

➤ *If the estimator carefully thought through what it would be like to physically apply the roof during preliminary site visits and the estimating process, the bid will lean toward accuracy, and the pre-planning should be very helpful.*

➤ *Before full mobilization begins, the project manager should go through the pre-planning again and organize the most mundane details with care. For example, where stockpiling will take place and how it will be handled.*

➤ *Do not forget the protection of all people and property during mobilization. Project security and safety must be a continual part of a company's policy.*

- A thorough mobilization meeting can iron out many of the logistics required for working together on the project.
- The mobilization gives everyone a chance to work through how they will approach the project now that they are at the site and can understand the logistics more clearly.

Delegating Responsibility

- Being at the site can make delegating authority for each task clearer than it was during the earlier meetings.
- Make certain that any major changes at this meeting are documented and signed or initialed.
- The mobilization meeting will give the owner the first chance to see the items you have discussed.
- The owner can see the importance of a maintenance contract and the advantages and disadvantages of the roof that has been chosen.
- The mobilization meeting is an excellent time to sell extras, such as a maintenance contract.
- This is a good time to get to know the architects in business in your market area.
- Have a thorough discussion with the architect about all details and complex areas.

fastfacts

➤ *Check closely for any deviations from the plans that have occurred during the construction that has taken place before your mobilization.*

➤ *Be sure to include all phasing where the trades overlap in the mobilization discussions. For example, clarify how the sheet metal sub and the framer will phase with your roofing work.*

➤ *Be certain to use the mobilization meeting to discuss all details that require overlapping of the trades carefully.*

➤ *If the owner does not attend a meeting, make certain that all important information is communicated.*

- For any specialty roofs, have the manufacturer's representative present at this meeting.
- After the mobilization meetings, the schedule should be reviewed carefully (Figure 2.1).

EQUIPMENT FOR THE JOB

There will be jobs that come along where large equipment such as kettles, sweepers, generators, and lifting devices would speed up the work.

Investing In Equipment

It is important to purchase equipment that will reduce labor considerably. However, timing is important. Reviewing your cash flow over a long period of time can help forecast where the peaks and valleys tend to appear in your profits.

Before major capital outlays are made, a discussion with your accountant can be a big help. For more information about the business of roofing, refer to McGraw-Hill's *Roofing Handbook*.

- If you do land a job that will show a considerably increased profit from using a piece of equipment, it is the perfect time to buy.
- Forecasting and leveling can be accomplished with computers to help the roofer make wise decisions about when to make major outlays of cash.
- If your records show that the period when you want to buy a piece of equipment is a time of year when cash is typically strong, and your payables and receivables are in good shape, it may be time to buy this piece of equipment.
- Rent equipment from several different manufacturers for evaluation by your field personnel.
- Discuss the benefits of leasing, and the depreciation of an outright purchase with your accountant.
- When you purchase a piece of heavy equipment, send out an announcement to all of the contractors, architects, engineers, and building owners who use this type of roof. Tell them that you are now better equipped to service their needs.

CHECKLIST FOR MOBILIZATION MEETING		
✔	ITEM	NOTES
	The final questions about details have been worked out.	
	All new contingencies related to the condition of the roof have been discussed.	
	All pick-work required of other trades has been reviewed with them.	
	Flashing and all roof penetrations have been discussed.	
	All questions related to specified products have finalized.	
	All equipment to be mounted on the finished roof has been discussed.	
	Any new potential for delays: access, weather, materials acquisition, overlapping of trades, site access, framing repair, etc., have now been worked through.	
	All staffing questions have been resolved.	
	All newly arisen refuse disposal issues, including environmental and recycling, have been addressed.	
	New opportunities to sell extras and change orders have been addressed with the owner and/or the architect.	
	Safety measures have been reviewed in detail.	

FIGURE 2.1   Mobilization meeting checklist.

- When introducing new gear, follow the manufacturer's operating instructions and warranty information at all times.
- Use approved safety precautions, safety equipment, and worker education to keep the jobsite safe and productive.

Hand Tools

- A crew's hand tools are among the most important equipment on any roofing job.
- Professional roofing mechanics keep their hand tools clean and ready to use whenever they are on the job.

The basic tools are:
- A nail bag with proper belt and hatchet holder.
- A double-knife sheath attached to belt.
- A utility knife with replaceable blades.
- A roofer's hatchet, depending on the type of roofing.
- A roofer's all-purpose utility bar.
- A set of tin snips.
- A chalk line and chalk.
- A wide bladed measuring tape.
- A roofer's tool belt or container for carrying tools.

- Knee and hip pads are also important.
- Wearing kneepads lessens job stress and makes working much easier.

fastfacts

➤ *Caution: When using the nail stripper (Figure 2.2), remember that it hangs from your chest when you're on a roof and at belt height when you're on a sidewall.*

➤ *Be sure to take the stripper off when you roll out felt. It can get in the way and pose a safety hazard.*

- For protection on very steep roofs, use hip pads.
- Some types of roof coatings require special hand tools.
- See the specific roof installation sections for more information.
- An example of specific hand tools is a nail stripper.
- A stripper can increase the production rate up to 50 percent on jobs that require the installation of strip asphalt shingles and three-tab asphalt shingles.

Power Tools

- Before deciding what power equipment to mobilize on the site, be sure to consider that the power supply must be sufficient to get the job done, with ample excess for motors pulling amps at start-up and unforeseen situations.

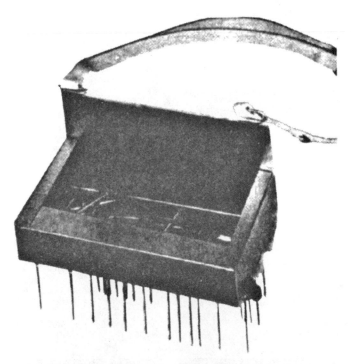

FIGURE 2.2 Typical nail stripper.

fastfacts

➤ *Caution: Utility knives are a must for roofers installing built-up roofing (BUR) systems, but they are considered a "don't" by many manufacturers of sheet membrane systems.*

➤ *A utility knife can cut through the second piece of a membrane and cause an instant leak. Because of this, shears or scissors are the preferred tool for single-ply systems.*

Generators

- Generators must be reasonably efficient in terms of fuel consumption and overall operating costs, and must be portable.

- The gas engine is the number one power source on roofing jobs because of its low purchase price, reasonable operating efficiency, and portability.

- In addition, tools such as drills, screw guns, circular saws, and other small hand tools, require an electrical power source.

- Before you buy a generator (Figure 2.3), figure out the maximum power you need.

- Look at which tools, or combinations of tools, are used at any given time.

- Add up the wattage of the tools used on jobs to arrive at your wattage requirements.

- Most generators are rated in watts (1000 watts = 1 kilowatt).

- If an electric tool is not rated in watts, wattage can be figured by multiplying volts times amperes (watts = volts x amperes).

- Another way to decide which generator to buy is by renting several types to try them out.

- Generators are available where other types of construction and industrial machinery are offered for rent.

fastfacts

➤ *Gas or diesel powered generators are necessary if you frequently encounter problems with electrical supply at the jobsite.*

➤ *Study how much labor is wasted searching for power supplies, running cords, throwing breakers, etc.*

➤ *If your crews need electric tools on the average jobsite, generators can pay for themselves quickly.*

Pneumatic Tools

- Roofers also use pneumatic power, especially when involved in asphalt shingle applications.

- While pneumatic power can be used for many tasks performed by electric tools, air power is used primarily for nailing and stapling (Figure 2.4).

- Economic conditions play an important role in any roofer's choice of fastening methods.

- A modern, lightweight portable compressor and pneumatic nailer, or stapler outfit, are fairly inexpensive.

- Air equipment can triple production when compared to typical hand application.

- Installation speed quickly translates into labor and the ability to handle more projects.

- Savings and more completed jobs enable the roofer to recoup the initial investment in pneumatic equipment quickly.

Special Equipment

- BUR and ethylene propylene diene terpolymer (EPDM) applications require special machinery and equipment, such as hot luggers (Figure 2.5), kettles (Figure 2.6), roofing torches, gravel spreaders (Figure 2.7), and tankers.

- Single-ply installation and roof tiles also require specialized tools, and tools that have general uses, such as wet-roof equipment.

FIGURE 2.3 MGH 10,000 generator; Master Distributors. *(www.mastergenerators.com).*

- The decision to buy gear should be made with care.
- Throwing money at tools randomly can lead to a rapid drain of capital.
- Study the market and make sure there is potential for an on-going cash flow before pulling out your checkbook.

DEWATERING DEVICES

- Sometimes the only practical way to get rid of the wet stuff is with the old-fashioned broom and mop routine.
- It is not easy to do, but it is a well-known fact that it works.
- On small jobs or tight spaces, using a broom and mop is the only way, short of evaporation, to get rid of water.

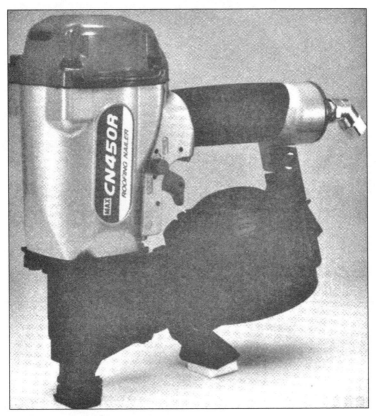

FIGURE 2.4 Max CN450R pneumatic nailer. *(maxusajdominice@cs.com)*

- When possible, it makes sense to mechanize water removal.
- The types of water-removal equipment currently available can be placed into three basic categories: dryers, blowers, and pumps.

Dryers

- Roof dryers are relative newcomers.
- The principle is simple and effective for drying up dew or shallow spots of water.

FIGURE 2.5 55-gallon hi-lo lugger; Reeves Roofing Equipment Co. *(www.reevesequipment.com /EquipCatalog/index.htm)*

- The dryer rapidly evaporates the water.
- Obviously, great care should be used with an open-flame device on the rooftop to assure that nothing or no one is burned.
- A later version of a roof dryer is a cart-mounted blower that uses a small engine to spread the heat produced by a Light Propane (LP) gas-fired burner.
- The blower can move large volumes of water, and the super-heated air serves to evaporate the dampness that remains.
- Care must be used to keep the blower moving while the torch is lit because the heat is intense enough to ignite materials.

FIGURE 2.6 450-gallon pump kettle with afterburner/safety loader system; Reeves Roofing Equipment Co. (*www.reevesequipment.com/EquipCatalog/index.htm*)

FIGURE 2.7 Hot process (left) and cold process (right) powered gravel spreaders; Reeves Roofing Equipment Co. (*www.reevesequipment.com/EquipCatalog/index.htm*)

fastfacts

➤ *If you encounter small jobs with tight spaces, or any situation where brooms, mops, and pails are the only choice for dewatering, schedule one or two men at the jobsite ahead of the rest of the crew. This eliminates time wasted waiting for the deck to dry.*

➤ *Do not use a blower for small jobs, when the blown water and debris might fall onto parked cars, or in any areas where people might be present.*

Blowers

- Blowers, also called air brooms, are useful for getting rid of shallow, ponded water.

- Available in backpack and cart-mounted configurations, air brooms do a more effective job than hand brooms, with more speed and less labor.

- Backpack models are the most popular because of their lightweight, low cost, and portability.

- In the recent past, small two-cycle engines and blowers have been greatly improved, while the purchase price has gone down.

- Look for a commercial-industrial model with a long blowpipe that easily reaches the deck surface.

- If you anticipate that you will use the air broom more than two hours per day, consider a cart-mounted model.

- Cart-mounted units are easier to operate over extended periods of time, and come with more horsepower and air-handling volume since weight is not a consideration.

Pumps

- The fastest and most productive way to get rid of large amounts of water is to pump it off.

- Several dewatering pump packages have been on the market for many years.

- The most popular package uses a marine-type, gas-operated, one inch pump with an automatic shut-off.

- These pumps can be operated unattended, and can move a respectable volume of water—approximately 2000 gallons per hour (gph).

- When buying, include a strainer screen at the suction end of these and other pumps. Without the strainer, debris can enter and damage the pump.

- Several pump packages offer specific advantages.

- At the low end of the cost scale is a hand-operated diaphragm pump that is used to start a siphon.

- The volume and speed are much lower than those of powered pumps (around 200 to 600 gph), but the cost of maintenance and upkeep is low.

- A small two-cycle engine, similar to the engine on the backpack blower, also powers a centrifugal pump.

- The price tag is low and the volume of water moved is respectable, at approximately 1750 gph.

- Hydraulically powered roof-dewatering pumps are available.

- The cost of the pump itself is quite nominal, given its production capabilities (approximately 6000 gph).

REROOFING EQUIPMENT

The roofing business can be very solid due to the immense number of retrofit opportunities. Reroofing is a mainstay for a large portion of the roofing industry, with virtually all roofers involved to some extent. With good management, reroofing can become a cash flow.

Powered Roof Removers

- Roof removers are very effective and relatively easy to operate.

- Powered roof removers (Figure 2.8) have a small gasoline engine that propels the unit and operates the flat blade on the front, which is from 12 to 24 inches wide.

FIGURE 2.8   Powered roof ripper, Reeves Roofing Equipment Co.
(www.reevesequipment.com/EquipCatalog/index.htm)

- The blade oscillates front to back, slicing through the membrane as the machine moves forward.
- In general, it takes a crew of four to eight people to keep up with one person operating the roof remover.

Powered Roof Cutters

- Designed for use on BUR roofs, the powered roof cutter (Figures 2.9 and 2.10) can speed production on virtually all flat decks.
- Weighing 200 pounds or less, the powered roof cutter cuts up the old membrane into sections that are easy to tear off and handle.
- It can be used ahead of a powered roof remover, and makes a nice even cut for tie-ins at the end of the day.
- This popular machine has been around for many years, and is available in models with 12 and 14 inch blades.

FIGURE 2.9 Single-blade roof cutter, Reeves Roofing Equipment Co.
(www.reevesequipment.com/EquipCatalog/index.htm)

FIGURE 2.10 Dustmaster Ultra Cutter, Garlock Equipment Company.
(info@garlockequip.com)

Powered Gravel Removers

- There is no doubt that a quality gravel remover (Figure 2.11) saves labor and dollars every time out.

- The small patch-scratcher class of gravel removers cut a path from 4 to 8 inches wide and are useful for small patches, and for scraping narrow paths on long runs, such as expansion joints or wall flashings.

- The full-size gravel removers cut a path wide enough to do entire roof surfaces, and are available in a wide range of sizes, from 12 to 24 inches.

- They also come with chimney shrouds and snorkel-type filters similar to those on roof cutters.

- Powered gravel removers come in a variety of power and blade configurations.

- To determine the best model for your needs, ask roofers in your area which one works best for them.

- The effectiveness of a blade design or cutting head depends on the type of aggregate used and the climate.

Powered Sweepers

- Today's powered sweepers (Figure 2.12) are available in several configurations.

- For example, there are models that feature 5-horsepower (hp) engines and 36 inch brooms.

- One model might utilize a belt drive to operate the wheels and broom, while another model might have a gear drive to the wheels and a belt drive to the broom.

- Most gas-powered sweepers are self-propelled and simple to operate.

- The right broom for the job is directly related to the task at hand.

Vacuum Systems

- Introduced in the mid-1970s, vacuum systems were designed to efficiently, quickly, and cleanly remove gravel from a roof surface and convey it off the building to a container on the ground.

FIGURE 2.11 Rocker powered gravel remover, Garlock Equipment Company. *(info@garlockequip.com)*

FIGURE 2.12 Sweeper 86, Garlock Equipment Company. *(info@garlockequip.com)*

- A full sized roof vacuum takes only a two or three person crew and, without dust, can remove gravel to the dump truck at a rate of 150 squares a day.

- Roof vacs are at their best when the jobsite conditions are at their worst.

- When gravel has to be removed from tall buildings that are cut up and inaccessible, a roof vac reduces logistical problems tremendously.

- The system utilizes a power unit, usually a 60 to 100 hp diesel engine, which drives a blower that creates a vacuum in a container or hopper.

- The 4 inch hoses extend to the rooftop, where the operator uses a pickup wand to vacuum up the loose gravel.

- Most vacs handle water as well, and most full-size vacs feature a reverse blower that conveys gravel back up to the roof.

- This method is not as productive or as efficient as some other hoisting/conveying methods, but it can return gravel to the rooftop efficiently at some jobsites.

Trash Chutes

- Reusable trash chutes are made of high-density polyethylene or other durable materials, and come in 4 foot sections that are 30 inches in diameter.

- Trash chutes are shipped and stored flat, then bolted together to make a cylinder.

- Trash chutes are used primarily for high-rise work, but they are also convenient at lower levels.

- The use of trash chutes can eliminate costly grounds cleanup and avoid damage to trucks, cabs, and windshields.

LADDERS

- Ladders have multiple uses in roofing operations.

- Roofers commonly use various types of ladders made from both aluminum and wood.

- Designs include single, straight, extension, and pushup ladders.

- A stepladder is a self-supporting, portable ladder that opens out, like a sawhorse.
- The maximum permissible height for a stepladder is 16 feet.
- Only use stepladders when they are fully opened, and do not use them as regular working platforms.

Roof Ladder

- A roof or chicken ladder (Figure 2.13) is used on steeply pitched roofs.
- Basically, it is a plank that has evenly spaced cleats, and is designed to accommodate the roofer's movements, but not materials.
- The crawl board plank should be at least 10 inches wide and 1 inch thick.
- The cleats should be a minimum of 1 inch by 11 inches and equal in length to the width of the crawl board.
- Space the cleats equally and not more than 2 feet apart.
- Drive the nails through the cleats and clinch them on the underside of the crawl board.
- Secure the crawl board to the roof by a ridge hook or other effective means.

ROOF BRACKETS

- Adjust roof brackets to fit the pitch of the roof.
- Space the brackets a maximum of 10 feet (Figure 2.14) apart, and secure them with multiple nails capable of supporting twice the maximum anticipated load.
- All planks should be a minimum of 2 x 10, scaffold-grade lumber.
- Extend the planks a minimum of 6 inches, and a maximum of 12 inches past the bracket, or overlap a minimum of 12 inches with both planks supported by the bracket.
- Either the roof edge should be protected with a 36 inch parapet, standard guardrail, and a scaffold platform, or the crew should use safety belts, lanyards, and lifelines.

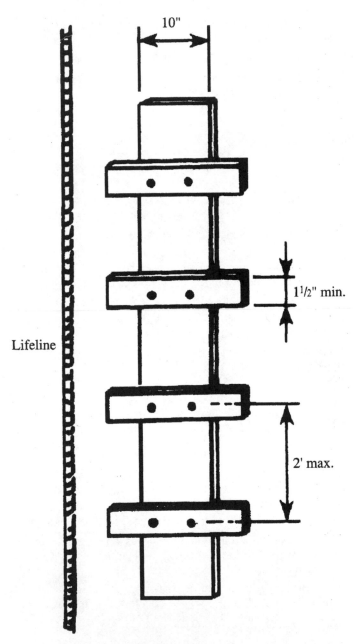

FIGURE 2.13   A roof or chicken ladder.

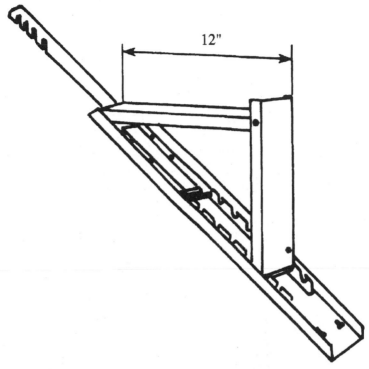

FIGURE 2.14 Roof bracket used on steep-slope roofs.

LIFELINE

- An adjustable roof seat, resting on self-gripping cleats, can be useful on steeply pitched roofs.

SCAFFOLDS

- Scaffolding consists of elevated platforms that rest on rigid supports.
- Scaffolds must be strong and rigid enough to safely support workers, tools, and materials.
- Scaffolds can be made of wood or metal.
- Wood staging is normally built on the jobsite.

- Metal scaffolds are manufactured in a variety of styles.
- Most of the scaffolding used today is manufactured metal.
- The scaffolds are easily assembled and dismantled.
- Many designs are available, including set tubular metal and rolling mobile. Figure 2.15 shows a tubular stationary scaffold unit.
- Scaffolds and lifts must be used in accordance with all local, state, and federal Occupational Health and Safety Administration (OSHA) laws and regulatory codes.
- Because of the high cost of warehousing and labor, it is cost-effective to subcontract scaffolding.
- This allows crews to move quickly on the roofing, making the job more profitable.
- Bid the scaffold subcontractor into the job, along with any miscellaneous costs you will accrue, and add a profit to the line item.

HOISTS

- A hoist of some sort is usually used to lift the roofing material onto the roof.
- A variety of roof-mounted hoists, truck hoists, telescopic hoists, trolley hoists, scissor-lift trucks, and cranes are available, one of which will be appropriate depending on the situation.

Roof-Mounted Hoists

- A roof-mounted hoist is, as the name implies, set up on and operated from the roof.
- The hoist framework supports the winch and power drive, and a wire cable extends from the roof to lift the materials off the ground.
- Accessories at the cable end adapt the hoist to the different roofing materials.
- These accessories include gravel buckets with dump gates for discharging gravel, hoisting forks for lifting full pallets of insulation or rolls, hoist slings for lifting long, heavy rolls of single-ply sheet goods, and platforms for miscellaneous materials or for lowering torn-off waste (Figure 2.16).

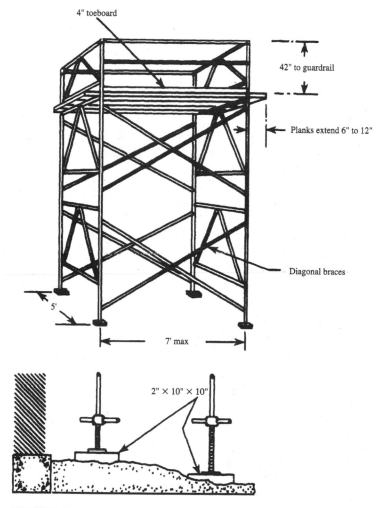

FIGURE 2.15 Typical tubular stationary unit and how it is secured.

- Roof-mounted hoists come in three basic styles: beam, swing-beam, and platform.
- Typically, the beam hoist can lift loads up to 300 pounds, while monorail or swing-beam hoists are capable of lifting loads of up to one ton in weight (Figure 2.17).

- Jobsite factors, such as setup space, material flow, parapet walls, and the sheer bulk of the materials, determine the best hoist.

- As with all major purchases, try renting first, track labor savings, and choose carefully.

Track Hoists

- Track or platform hoists come in a variety of sizes with several power and attachment options.

FIGURE 2.16 Typical platform hoist.

FIGURE 2.17 A swing beam hoist in use.

- There are many distinct advantages to using track hoists, especially the low initial cost.
- These rigs are lightweight and portable enough to be dismantled and carried easily in a pickup truck.
- They can be moved around the building as work progresses, a feature that is especially helpful when using scaffolding.
- Steep-slope and low-slope roofs are no problem for track hoists.
- The basic rig consists of a track, which is an aluminum frame that resembles, but should not be confused with, a ladder.
- Tracks come in several sizes and can be joined with other track sections to assemble the proper amount of reach.
- Track sections are joined by splice plates that slide into the sides of the tracks.
- This assures a snug fit and straight track.

Telescopic Hoists

- Another ground-to-roof hoist is the trailer-mounted, towable, telescoping hoist.
- This type of hoist is capable of extending up to 130 feet, and can carry loads that weigh from 400 to more than 1000 pounds.
- This type of hoist has actually been around for many years, but advances in technology have added greater height capabilities and lifting capacities.
- This versatile, easily transported hoist requires minimal setup space. These features are the reason telescoping hoists are popular.

Trolley Hoists

- The trolley track hoist was developed specifically for the BUR contractor.
- It generally has a capacity of approximately 400 pounds, though some can move more weight.
- The hoist is counterbalanced with iron or concrete weights set on a mechanically attached metal pan or tray.
- The counterbalance weight is at least half of the weight of the material being hoisted. Do not exceed the manufacturer's recommended weight limit.
- The aerial platform, or high lift, provides a means of working on a roof or providing transportation to the rooftop.
- Trolley hoists are driven by gasoline, diesel, or electric-powered motors.
- Most hoists of this type have a dual braking system. Another safety feature, required by OSHA, is 4-foot gates on either side of the trolley hoist.

Scissor-Lift Trucks

- Also known as high lifts, this type of hoist incorporates the truck used to haul materials and the lifting equipment.
- The body of the truck uses hydraulics to lift to the rooftop and then the roofing materials are walked or carried by wheelbarrow directly onto the roof.

fastfacts

Drawbacks to a high lift are that the reach is limited to approximately 18 feet and the truck has to be driven all the way to the building, which can be a problem on some job sites.

- Scissor-lift trucks are versatile tools for roofers, particularly because they can be used for both tear-offs and BURs.
- Most scissor-lift trucks have the capability of dumping, making this a very nice piece of equipment.
- Set up the unit on level ground with its outriggers down and unloaded evenly.
- For the residential shingler who does reroofing and mostly one story work, this rig is hard to beat.
- The scissor-lift truck is also designed to work with a belt conveyor.

Cranes

- Cranes are another popular piece of lifting equipment.
- If you have frequent jobs that require a crane rental, it might be more economical for you to purchase or lease this piece of equipment.
- A variety of attachments are available, from tear-off boxes and gravel buckets, to hoisting forks and single-ply slings.
- Keep in mind that, of all the equipment available for moving materials, the crane option is probably the most expensive initial investment.
- Be sure to try rentals and explore leasing before buying.

CONVEYORS

- Conveyors can be used to transport all types of material to the construction worker on the roof.

- Conveyors are often misused like the ladder and hoist. However, they should not be used to transport equipment or workers.
- Belt-flight and chain-driven conveyors provide more protection on large, flat, BUR roofing projects.
- They have evolved to meet the needs of most new single-ply roofing systems.
- One advantage to the conveyor is its ability to continuously move material to the top without having to stop and return to the ground.
- Another advantage is the conveyor's ability to move gravel at a rate that outpaces the best of crews.

Truck-Mounted Conveyors

- Truck-mounted conveyors offer several advantages over some other methods of getting materials to the job and onto the roof.
- When the conveyor unit is mounted on the truck, there is essentially one piece of equipment that moves the material from warehouse to jobsite and then onto the roof.
- This eliminates the need for other tow-behind equipment and saves setup time.
- A continuous-drive conveyor, even moving at relatively slow speeds, can deliver materials at a rate faster than most crews can load and unload them, often up to 720 bundles an hour.

MAINTAINING TOOLS AND EQUIPMENT

- Preventive maintenance is less expensive and more sensible than crisis maintenance, but preventive maintenance lacks the urgency of purpose.
- Preventive maintenance is not urgent unless it is made urgent by diligent company practice.
- A sound maintenance program is not always the easiest thing to get started, but once going it can pay some nice dividends.
- Put one person in charge of your maintenance program, and work with him or her to get the job done.
- Follow up to see that the system works correctly.

- Encourage all mechanics and operators to report any malfunctioning equipment immediately.

- Establish a reporting method for everyone to use.

- The best way to accomplish this is to enforce a policy that all equipment be kept clean and looking good.

- When equipment is kept clean and nice looking, employees take the attitude that the equipment is valuable to their company and that care must be taken when it is used.

- Cleaning also helps certain pieces of equipment perform better technically.

- Equipment that appears cared for reflects the entire attitude of the company: hit-and-miss, or professional.

SETTING UP THE SITE

- All of the various types of roof work have unique mobilization requirements.

- Review the section in *Roofing Instant Answers* that is specific to the type of roof installation you are mobilizing. There is information in each installation section that is helpful with mobilization.

- The following generic concepts apply to all installations:
  - Protect all adjacent structures.
  - Protect all property on the site.
  - Protect all trades people who are working on the site.
  - Install all safety devices and requirements before any work commences.
  - See Chapter 3 for more details on site safety.
  - Have definite safety procedures in place for your company.
  - Have a person on site that takes care of safety.
  - Post all critical telephone numbers: fire, paramedics, environmental emergencies, etc., where they can be quickly accessed.
  - Post all safety information in English, as well as any other languages required to notify all crew members.
  - Keep fire extinguishers serviced.
  - Have readily-available, freshly stocked first aid kits of ample nature for any emergency.

- Stockpile materials where they can be readily accessed for transfer to the roof deck.
- Protect all roofing materials as they are stockpiled.
- Monitor the condition of all roofing materials regularly if they will be on site for lengthy periods of times.
- Provide parking for all crew members.
- Be aware of traffic control required by the local municipality.
- Secure equipment so that it will not create a hazard after the workday.
- Be sure to remember to provide space for bringing in heavy equipment.
- Remember to provide space for debris removing equipment.
- Double-check all environmental requirements.

MANAGING INVENTORY

Every business needs to keep a certain amount of material on hand. There was a time when many businesses required large amounts of warehouse space and storage yards, which meant payments and maintenance for large parcels of real estate. However, in recent years, companies have changed to methods that allow the materials to be called for as they are needed.

- Direct materials prices are only one part of the cost picture.
- Handling costs, storage costs, shipping costs, and losses from theft or obsolescence are all factors that can contribute to shrinking profits.
- The technology for effectively dealing with inventory costs and increasing profits by reducing the amount of stored inventory is now readily available.
- Bring your vendors into the technology loop, size up what automated services they offer, and how the services work with the software you are going to buy.
- Work with suppliers and your software vendors and consultants to establish an inventory control system based on when you use the product.
- This "just-in-time" system refers to the ordering of materials just in time for their use on the job.

- More and more wholesale roofing suppliers are equipped with crane or conveyor trucks with which they can deliver materials to the jobsite and directly onto the roof.

- This is particularly true of suppliers delivering slate or tile products.

- Naturally, there are some materials that every roofer needs to keep in stock, such as an assortment of fasteners and felts.

- On almost every roofing job, whether new construction or re-roofing, there are products like mastics and felt that will be needed.

- Buying commonly used products in quantity and having them handy can save in purchasing costs, downtime, and shopping time on payroll.

- You can also use these materials from the warehouse to start jobs and keep them underway until the other materials are delivered by the supplier.

- Just-in-time materials management is a direct benefit of technology.

- It should be remembered that just-in-time orders can loop right into your system at the beginning of a job during estimating.

- Once the job is awarded, it is a simple matter for software to translate the estimate sheet into a materials list.

- This list then can be emailed to vendors for bidding, directly from the computers in the office or a laptop at the site.

- Using the computer's integrated software, you can create a record—or file of materials on order—that includes the job name, job number, date ordered, supplier name, vendor number, and other required information.

- Furthermore, the felts and fasteners used on a particular job can be automatically removed from inventory by the computer and re-ordered or placed on an order sheet.

- These software programs can, with one instruction from the system operator, condense several time consuming tasks into a one-step process that orders materials upon request, updates accounts payable, and calculates the vendor's discount schedule.

- The just-in-time inventory philosophy keeps obsolete products off warehouse shelves.

- With so many colors, styles, and types of roofing materials available, one shingle might be popular today and out-of-style tomorrow.

- The longer material stays on your shelves, the more money you lose.

- The just-in-time policy eliminates the need to invest in large quantities of stock that might not immediately turn into profits.

- By reducing the volume of materials kept in physical inventory and having products delivered directly to the jobsite when needed, roofing contractors can reduce insurance, handling, labor, and carrying costs, which translates to an improved bottom line.

JOBSITE COMMUNICATIONS

Communication is one of the most important parts of being an effective, competitive roofing contractor. All communications— between the owner, architect, suppliers, manufacturers, and the office—must be considered for all job sites.

Electronic Equipment as Important Tools

- For very simple work, a cell phone can fill the bill.
- But do not be afraid to invest in communications technology.
- Don't be stuck in the mindset of thinking that equipment such as nail guns are the only tools.
- Think of electronic equipment like computers and faxes, as tools.
- Instant information is a must. Today, there is no way around considering electronic communications equipment absolutely necessary.
- Just-in-time inventory is a perfect example of just how important it is to think of electronic equipment as valuable tools.
- Without computers and fast signal lines, just-in-time inventory would be impossible.

Mobilizing Electronic Equipment

- There are many tasks that can be easily handled electronically:
  - Phone installation can be replaced by using cell phones.
  - Computer work can be handled on-site with laptops.
  - Faxes can be sent and received using a laptop.
  - E-mail can be transferred via cell phones.
- Materials can be dispatched by email, which tends to keep the lead man off the telephone.

fastfacts

➤ *Do not hesitate when it is time to invest in communications equipment. Building owners and architects are not impressed by contractors with whom it is hard to communicate. Being readily available is very good marketing.*

➤ *Make your communications systems simple, thorough, and fast.*

➤ *Do not spend money that you cannot afford, but never be "penny wise and pound foolish" about your electronic capabilities.*

- Change orders can be taken care of via e-mail.
- Any detail drawings can be attached in an e-mail, which can save considerable down time when the crew is waiting on the architect.
- Equipment rentals can be taken care of with e-mail.
- Payroll can be done on a laptop, and e-mailed directly to the office.
- Payroll can even be dispatched to the employees' banks from the office, via electronic transfers.
- Schedules can be updated on a regular basis.
- The project manager's truck can become an electronic office.

WASTE DISPOSAL

- Waste disposal is included here because it has become a very important part of planning any roofing job.
- The site must be examined closely.
- If it is new work, and the roof is a part of a larger general contract, then waste disposal must be discussed with the principal and the general contractor.

- If it is a reroof, it is very important to plan how the material will be evacuated from the roof deck, and to know the nature of the material and how it will be evacuated from the site.

ENVIRONMENTAL ISSUES

- Environmental concerns are an area of increasing importance to the roofing industry.
- Standard procedure needs to be analyzed and updated to comply with today's ever-changing social and legal climate.
- Disposing of old roofing and packaging material may not be a simple matter of trucking it to the nearest landfill.
- Many landfills are close to capacity and some municipalities and cities allow only specific materials to be dumped in their landfills.
- This means that you might end up sorting debris for disposal at two or more landfill sites, then driving to a landfill that takes the material.
- The cost of disposal is a significant expense for any roofing project.
- Consider too, the additional costs involved with certain types of debris that might not be allowed in landfills anywhere near the vicinity of the job, the entire state, or even the area of the country. The cost of hauling debris to one or more landfills, if the local landfill does not accommodate the volume of debris generated by a major reroofing job, must be considered.
- The resulting cost of either hauling this debris to another site, or arranging to have it buried by itself, can devastate your profit margin.
- In order to limit expenses, roofing contractors are adopting ecological policies.
- Any reroofing procedure that can be altered in order to reduce costs is being scrutinized.

CHECKLISTS FOR WASTE DISPOSAL AND ECOLOGICAL POLICIES

The checklists in Figures 2.18 and 2.19 can be used to tune in your firm to environmental issues and, through the effort, save money and add environmental achievements to your marketing documents. They can be expanded as you bring your firm to the forefront in environmental practices, as many large firms have been doing for years.

- Do not adapt an attitude of foot dragging and complaining about environmental constraints.

- Use your company's deep environmental knowledge and policies as marketing tools.

- Remember, markets for environmentally friendly roofing can become good, strong cash flow with little competition.

- With dumps close to capacity, and some municipalities and cities allowing only specific materials to be disposed of in their landfills, you might end up sorting debris for disposal at two or more sites.

- Use challenges proactively rather than complaining, and make everyone aware that your company is completely abreast of this information.

CHECKLIST FOR WASTE DISPOSAL		
✔ ITEM		NOTES
The cost of disposing the waste in a local landfill, if possible.		
Will debris from the job need to be sorted for disposal?		
If sorting is required, how many work hours are required?		
Can all debris be disposed of in normal fashion?		
If not, how is disposal accomplished?		
At what cost?		
What products cannot be disposed of in the local landfill?		
What are the locations of the landfills for materials that cannot be disposed of in the local landfill?		
Where are the landfills that will accept the items that cannot be disposed of locally?		
What is the trucking cost involved?		
What are the dump fees?		
How is the ownership of the disposed products designated?		
What other stipulations go with these sites?		

FIGURE 2.18 Waste disposal.

CHECKLIST FOR ECOLOGICAL POLICIES		
✔	ITEM	NOTES
	Recycle existing roofing materials, such as aluminum, copper, etc.	
	Reuse existing materials, such as insulation or gravel surfacing, on the same roof.	
	Reassess the existing roof system to determine if it can accommodate the new roof on top of it.	
	Can the materials packaging be incorporated into the roofing system?	
	How else can the existing roofing material be recycled?	
	At what cost?	
	What products cannot be recycled and must be sent to a landfill?	
	Review all materials with state and federal agencies for possible recycling compensation.	

FIGURE 2.19   Establishing ecological policies.

- Being up-to-date is very important because serious troubles can arise from the disposal of hazardous materials. You, your crew, or the principal may end up being responsible for the material even after it is in the disposal site.

- One way to improve profitability, while becoming more green, is to study, look for, and implement innovative methods for disposing of old roofs.

- Roofing materials, such as tile, that cannot be used on the existing roof might be of use to other contractors.

- Salvaging material and selling it through a broker or to another contractor eliminates disposal costs, increases revenues, and allows a fellow contractor to purchase materials at a discount.

- Contact the EPA for your region and the state environmental agency. There are many innovative programs underway, and you may find several that serve your needs and save, or even make, money for disposal.

- If treated in a pro-active manner, recycling and environmental matters can, and should, become profit boosters.

chapter 3

SAFETY MEANS SUCCESS

Safety does not come to the jobsite in the worker's toolbox. It has to be carefully cultivated. Whether the roof is to be covered in tiles or shingles, a good roofer knows that business overview, management, workmanship, skill, the right tools, proper loading equipment, and a watchful attitude go hand in hand to create a quality roofing job and a safe roofer.

BEGIN SAFETY SKILLS IN THE OFFICE

- Put training procedures into place.
- Start by developing your own individual training program.
- The training program is one portion of your overall safety program.
- Naturally, safety for the protection of the field crews is necessary, but the office staff also should be included.

PROTECTING THE WORKER

Considerable research has been directed toward improving personal protective equipment in the construction industry. No matter what the degree of sophistication of the safety equipment, it is effective only when used properly.

- All workers should wear proper protective equipment that is in good repair.
- Roofers working with flammable substances, climbing ladders, and handling materials should wear clothes that are comfortably snug, particularly around the neck, wrists, and ankles.
- There should be no loose cuffs, flaps, or strings.
- Machinery operators should not wear neckties, loose sleeves, rings, watches, or long hair, all of which can be caught in equipment and lead to injury.
- Other personal protective equipment should include the following:
  - Safety glasses, goggles, and plastic face shields when working with material that might become airborne.
  - Hardhats approved for potential exposures when there is any possible hazard from above.
  - Closed-cuff, chrome-tanned leather or insulated gloves in good condition and suited to the handling of roofing materials.
  - Safety shoes that meet nationally recognized standards.
  - Approved respiratory protective devices proper for the existing hazard (dust, fumes, smoke, vapors, mist, etc.).
  - Safety belts when working at elevated levels that are not protected by handrails or when working from suspended scaffolds.
- First aid kits and a Red Cross first aid manual always should be handy.
- When working with roofing chemicals, have these additional items available:
  - Gauze for burns and eye injuries.
  - Baking soda to neutralize acid.
  - Mineral oil, olive oil, or baby oil to help rinse burning materials from the eye.
  - Blankets for shock.

PHYSICAL SAFETY

- Every year, roofers suffer back injuries from lifting heavy materials incorrectly.
- Besides being extremely painful, back injuries can be the start of other recurrent back problems.
- Lifting improperly can cause muscles to tear and/or ligaments to rupture.
- A muscle tear is exactly what it says, a tear in a muscle from excess strain or stretching.
- Ligament rupture occurs when the tissue breaks from the bone.
- To avoid back strain, stop any lifting if you cannot keep your hips under your upper body when you begin to stand up.
- The American Chiropractic Association (ACA) offers five lifting rules that a roofer should remember and use in order to help avoid back strain (Figure 3.2).
  - Plant feet 12 to 18 inches apart on solid ground in front of the object.
  - Squat down in front of the object and keep your back straight when lifting.
  - Lift with the muscles of your legs, thighs, arms, and shoulders, not the back.
  - Keep your arms close to the object and have a clear view of the path to be taken.
  - Never lift more than you can handle. Use a lifting device if the load is too heavy.

Guarding Against Falls

- Most roofers love the thrill of working high above the ground, but the thrill can be lost with a single slip.
- The risk of falling is significant and can be amplified by bad habits.

Climbing Ladders

- The ladder, an indispensable mainstay of every construction site, is often taken for granted.

JOB SAFETY CHECKLIST		
✔	ITEM	NOTES
	Safety glasses, goggles, and plastic face shields when working with material that might become airborne.	
	Hardhats approved for potential exposures when there is any possible hazard from above.	
	Closed-cuff, chrome-tanned leather, or insulated gloves in good condition and suited to the handling of roofing materials.	
	Safety shoes that meet nationally recognized standards.	
	Approved respiratory protective devices proper for the existing hazard (dust, fumes, smoke, vapors, mist, etc.).	
	Safety belts when working at elevated levels that are not protected by handrails, or when working from suspended scaffolds.	
	First aid kit and Red Cross first aid manual easily accessible.	
	Ample gauze for burns and eye injuries readily available.	
	Baking soda available to neutralize acids.	
	Easy access to mineral oil, olive oil, or baby oil to help rinse burning materials from the eye.	
	Blankets for shock available	

FIGURE 3.1 Job safety checklist.

- Perhaps the complacent feeling toward the ladder comes from its familiarity.
- It is, after all, the most widely used roofing tool.
- It is probably safe to say that most roofing-related OSHA citations involve the improper care and use of ladders (Figure 3.3).
- It is also safe to surmise that nearly all those citations, and the injuries that result from falls off ladders, could have been prevented with a healthy dose of common sense.
- An analysis by OSHA of accidents involving ladders reveals that the four principal causes of such accidents are:
  - Structural failure of the ladder itself.
  - Failure to secure the ladder at the top and/or bottom.
  - Ascending or descending improperly.
  - Carrying objects in hands while ascending or descending ladders.

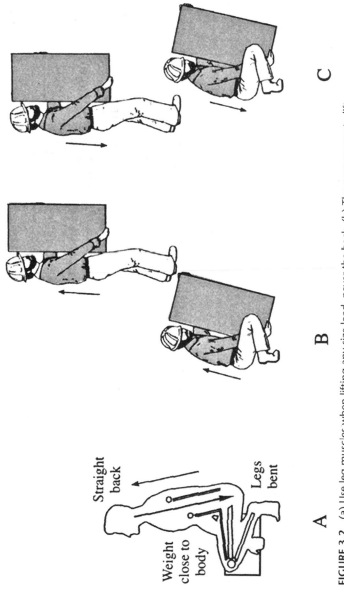

FIGURE 3.2 (a) Use leg muscles when lifting any size load, never the back. (b) The proper way to lift a shoulder-high load. (c) Put down the load as carefully as it was lifted.

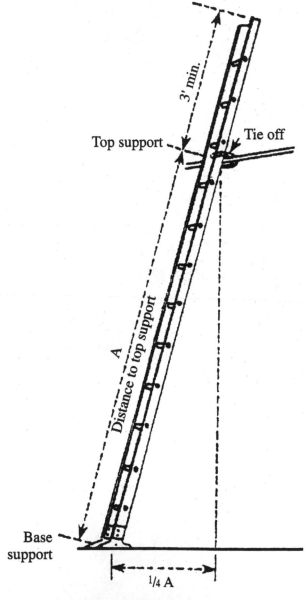

FIGURE 3.3 Make sure the ladder is about one foot away from the vertical support for every four feet of ladder height between the base support and the top support.

Using Scaffolding

- Scaffolding requires frequent safety checks.
- Any damaged scaffolding accessories, such as braces, brackets, trusses, screw legs, etc., must immediately be removed for repair or replacement.
- All scaffolding must have solid footing.
- OSHA requires that scaffolds must be engineered to carry at least four times the maximum intended load.
- Never support scaffold on unstable objects such as barrels, loose bricks, boxes, or planks.

Safeguarding Roof Edges

- Roof edges should be guarded as a standard practice for all quality roofing firms.
- Standard OSHA guardrails can be built with 2 x 4 lumber.
- Height is between 36 and 42 inches.
- Attach guardrails by use of counterweights, attachment to joists, or to the wall or the roof of the structure (Figures 3.4 through 3.6).
- Warning lines offer several advantages over guardrail systems.
- The warning line system consists of stanchions, and rope or wire guarding lines.
- Warning lines comply with OSHA standards.
- The OSHA standard requires that any flags or pennants be placed 34 to 39 inches from the roof surface.

fastfacts

Use the warning line system on all flat roofs, regardless of size, whenever mechanized roofing equipment is used.

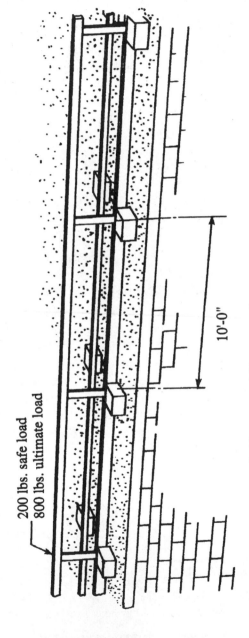

200 lbs. safe load
800 lbs. ultimate load

10'-0"

FIGURE 3.4 Guardrail secured with counterweights.

74

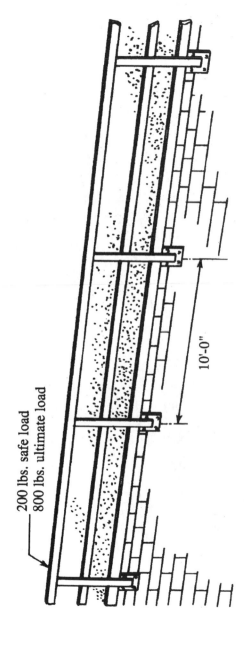

200 lbs. safe load
800 lbs. ultimate load

10'-0"

FIGURE 3.5 Guardrail attached to exterior wall.

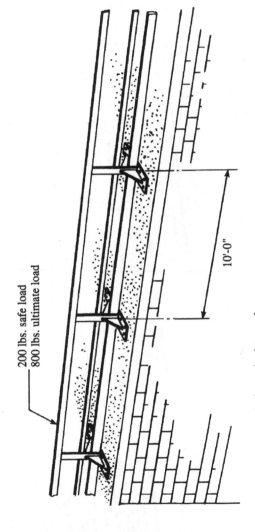

200 lbs. safe load
800 lbs. ultimate load

10'-0"

FIGURE 3.6 Guardrail attached to roof.

Barricading Open Roof Holes

- Holes are every bit as dangerous as falling over the edge of the roof deck.

- Many situations during the build-out of a structure require open holes in the roof deck.

- Obviously, holes through the deck are hazardous unless they are protected by some type of barricade.

- It is essential to guard all openings in roof surfaces with a standard barricade and toeboard on all exposed sides.

- A toeboard is a strip of wood or metal, 3 to 6 inches in height, that is placed along all exposed edges of any deck opening.

- Besides helping to keep a worker from falling, toeboards keep tools from being kicked off and injuring a worker below.

- Any opening that has a maximum gap of 3 feet or less can be covered with plywood or material that is strong enough to bear all intended stress.

- An opening that encounters heavy traffic needs extra reinforcement.

- As the size of the opening increases, add extra support to the covering.

- If the opening is too big to cover, erect OSHA standard guardrails.

- Never place a board, or any makeshift bridge, over the opening.

- Bridges and boards only encourage people to walk over the hazardous area.

- Specialized barricade options for floor openings and hatchways include:

  - A hinged cover, with someone in constant attendance when the cover is open.

  - Portable railings on all exposed sides except at entrances to stairways.

  - Standard railings for temporary floor openings.

Installing Catch Platforms

- Catch platforms are installed at or near the roof perimeter.

- When attached below the roof (Figure 3.7), the catch platform allows complete freedom for roofing operations.

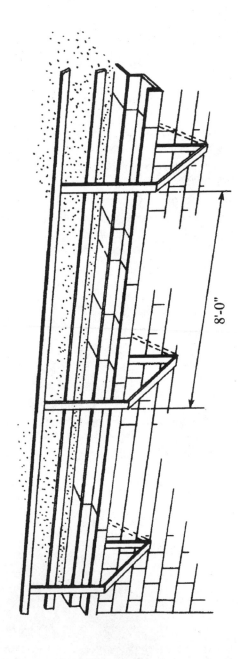

FIGURE 3.7 Perimeter catch platform.

- The platform can also serve as a work surface for perimeter work.
- The system is technically feasible for many buildings.
- Masonry veneers invite uncertain, and possibly dangerous, horizontal anchorage.
- The system is useless in situations where walls have high windows, or are constructed with lightweight panels.

The roof-attached catch platform system shown in Figure 3.8 is installed remote from the roof edge. This eliminates the use of lifelines. It can also serve as a work surface for perimeter operations. The catch platform shown in Figure 3.9 is hung over the roof edge and attached only at the roof deck level. This, in turn, eliminates the need to fasten it to the wall system. This catch platform scheme protects personnel in continuous roofing operations, but limits perimeter work. Lifelines or other safety systems must be used when installing, removing, and patching holes in the roofing system left by safety devices.

Curbed openings pose a threat too, though not as much as flush openings. Rail or cover these openings if they do not extend more than 36 inches. A roofer could accidentally back into a curb edge, and fall over it and into the opening if these areas are not properly protected.

Falls also occur when workers are setting panels. Keep walking on girts and purlins to a minimum when this work is being performed. Assure worker control over hoisted panels by using taglines. Also, check roof decking when it is placed to ensure that the panels are supported on all four sides.

CARING FOR INJURIES

- A minor slip on a roof can cause a serious accident.
- Eliminating the causes of slips is the main method a roofing contractor can employ to significantly reduce falls at the jobsite.
- Keep from moving anyone who falls from one level to another.
- Have him or her lie flat, even if he or she does not seem to be injured.
- Do not move the victim.
- Back injuries are often hidden, and any manipulation of the person can aggravate the problem and cause serious damage.

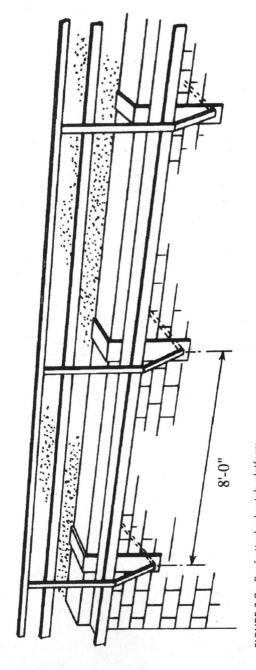

8'-0"

FIGURE 3.8 Roof-attached catch platform.

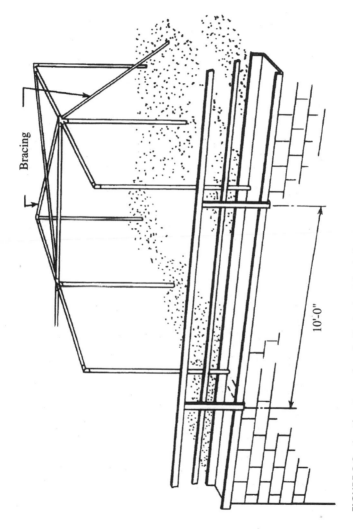

FIGURE 3.9 Another type of roof-attached catch platform.

Bracing

10'-0"

- Ideally, only trained medical professionals should move any worker who has fallen.
- If a worker falls into an awkward position, however, it might be necessary to slowly and gently straighten his or her back.
- Do this only if there are other hazards present, such as the injured person falling further or having trouble breathing.

STARTING A SAFETY PROGRAM

- Safety cannot be accomplished simply by writing a safety policy.
- Designating a safety officer, and performing safety training are a must.
- Also, follow-through is absolutely essential.
- The only thing that a company can do to enhance its employees' natural desire not to get hurt is to give safety as much emphasis as possible (Figures 3.10 and 3.11).

	SAFETY CHECKLIST
✔	ITEM
	All Jobs, Trucks, Warehouse, and Office
	Fire extinguisher-with service inspection-up to date.
	Safety glasses-in stock.
	All safety signage posted.
	Safety cans for flammable fuels-recently inspected.
	Industrial-grade first aid kits-onsite and in trucks, recently inspected and stocked.
	Flat Roofs
	No mechanical equipment or material within 6 ft of any edges of roofs or large openings.
	Warning lines or fall monitoring systems in place, checked, and working.
	Tear-offs
	Chute used for disposal at distances greater than 20 ft.
	Inspected safety glasses and dust masks available and worn by crews.
	Tear-off box roped off.
	Special Hazards
	Electrical lines shut off or roped off.
	Roof openings covered or roped off.
	Unsafe decking properly covered.
	Flammable vapors discharged.
	Radiation hazard on roof-see owner.
	Kettle
	Inspected
	Face shields and proper clothing worn by crew.
	Fire extinguisher-inspected and easily reached.
	Placement reviewed for safety.
	Propane storage secure and 15 ft from kettle.
	Guardrails at outlet area.
	Dangerous areas roped off.
	Proper signage in place.

FIGURE 3.10 Safety checklist (to post on roofs).

✔	SAFETY CHECKLIST
	ITEM
	All Jobs, Trucks, Warehouse, and Office
	Fire extinguisher-with service inspection-up to date
	Safety glasses-in stock
	All safety signage posted
	Safety cans for flammable fuels-recently inspected
	Industrial-grade first aid kits-onsite and in trucks, recently inspected and stocked
	Flat Roofs
	No mechanical equipment or material within 6 ft of any edges of roofs or large openings
	Warning lines or fall monitoring systems in place, checked, and working
	Tear-offs
	Chute used for disposal at distances greater than 20 ft.
	Inspected safety glasses and dust masks available and worn by crews
	Tear-off box roped off
	Special Hazards
	Electrical lines shut off or roped off
	Roof openings covered or roped off
	Unsafe decking properly covered
	Flammable vapors discharged
	Radiation hazard on roof-see owner
	Kettle
	Inspected
	Face shields and proper clothing worn by crew
	Fire extinguisher-inspected and easily reached
	Placement reviewed for safety
	Propane storage secure and 15 ft from kettle
	Guardrails at outlet area
	Dangerous areas roped off
	Proper signage in place

FIGURE 3.11 Safety checklist (to post on the ground).

- The first prerequisite for any safety program is a commitment from top management.
- Without this commitment, the safety program is going to be handicapped at best, and doomed to failure at worst.
- Management must understand that safety can enhance profitability.
- Safety can lower insurance premiums and reduce OSHA, Department of Transportation (DOT), and Environmental Protection Agency (EPA) citations and fines.
- It can improve morale and productivity.
- Safety can reduce substance abuse.
- It can keep a firm away from negative publicity.
- Safety programs can even be used as a marketing tool to separate your company from your less safe competition.
- Look closely at the type of work you do and the equipment you use.

- Different operations have different hazards.
- In addition to the information in this chapter, the National Roofing Contractors Association and OSHA can help answer questions about hazards, and how to protect your company and your employees.

chapter 4

MAINTENANCE, CALLBACKS, AND ANCILLARY BUSINESS

The smart roofer does not avoid callbacks or maintenance. First-rate roofing firms have very few callbacks because they work smart, not hard, and they turn maintenance and warranties into cash by writing service contracts.

- Working smart means planning carefully, hiring excellent crews, and doing it right the first time.

- Since construction projects are performed in the open, with ever changing project constraints, and at the mercy of the elements, roofing errors are inevitable.

- Aside from snafus that are beyond the control of the roofer, several things cause construction defects and callbacks:

  - Accepting jobs that are low-balled, with poor contracts, and for people who are not in control of their work.

  - Lack of careful planning.

  - Hurrying, rather than working steadily.

  - Not training or taking care of crews well

  - With diligent attention to the items listed above, callbacks can be slowed to a minimum.

  - The first-rate roofing company knows that the long life of a well-installed roof depends on maintenance.

GENERATING CASH FLOW FROM MISCELLANEOUS ROOF WORK

If you are serious about taking charge of your business, setting real-world goals for yourself, and proceeding with a day-by-day approach to furthering those goals, small, simple ancillary cash flows can be invaluable.

WARRANTIES AND MAINTENANCE CONTRACTS AS ADDITIONAL REVENUE SOURCES

Warranties

In the new millennium, things are going at a fast and furious pace. In these complex times, roofing contractors must often wear more than one hat. For example, when dealing with construction defects and workers' compensation, the business owner must be both a professional roofer and a bit of a lawyer.

- When working with warranties, contractors often find themselves caught between the manufacturer and the customer if there is a roof failure.

fastfacts

➤ *Callbacks will never be prevented completely, so the cost to the roofing company for callbacks must be factored into the overhead when estimating.*

➤ *Parallel businesses to your roofing company, such as carrying products, can be started simply and will enhance your current assets.*

➤ *We have addressed only a handful of the possibilities; if you want more, just open your eyes and start to watch for them.*

➤ *The chances are strong that you will find more sources of possible income than you could ever dream of pursuing.*

- Manufacturers' product warranties give the customer rights and protection against failure of the product for a given length of time and under certain conditions.

- Customers often look to the roofing contractor to uphold those rights.

- A condition is a provision in the warranty that can, unless met by the owner, lead to the denial of any claims, or even the cancellation of the warranty.

- Another common provision is that the warranty does not take effect until all bills for the installation of the materials have been paid to the manufacturer, suppliers, and contractor.

- Other conditions deal with misuse or negligence by the owner, or acts of vandalism that damage the roofing membrane.

- These are sometimes referred to as exclusions to the warranty. Persistent problems along these lines can be grounds for cancellation of a warranty.

- When a roof failure occurs, the owner looks to the contractor to honor the warranty.

- The contractor then looks to the manufacturer, who denies the warranty due to lack of payment. Guess who is sued.

- Make sure that you fully understand warranty terms and conditions, including the manufacturer's responsibilities, and any conditions that might exempt the manufacturer and supplier from responsibility and leave you holding the bag.

- Under commercial law, all products carry an implied warranty of merchantability by the manufacturer.

- This means that the product conforms to the ordinary purposes for which goods of such description are generally used.

- If the product is presented as a roofing membrane, it must function satisfactorily as a roofing membrane.

- Any use of this material other than as stipulated can be referred to as implied use.

- This is also known as a disclaimer. It is legal and can be upheld and enforced in court as long as it is in writing and in clear, conspicuous, or noticeable print.

- When you bid a job, make sure that you know what is warranted and what is not.

- Do not make any representations to the owner that items or situations are covered by the warranty when they are not.

fastfacts

➤ *In order to protect themselves from warranty claims resulting from a product failure under circumstances of implied use, manufacturers usually include a provision within warranties that states something to this effect: "The obligation contained in this guarantee is expressly in lieu of any other obligations, guarantees, or warranties, expressed or implied, including any implied warranty of merchantability or fitness for a particular purpose. In no event shall we be held liable for consequential or incidental damages of any kind."*

➤ *Another statement common in warranties is a statement of limitations, in which the manufacturer limits liability to an amount not to exceed the original cost of materials.*

• Roofing contractors often make it standard practice to have the owner sign a document that states that the owner has read and understands the warranties that accompany the materials.

• This precaution, while not eliminating the possibility that the owner will make a claim against the contractor for work that is not warranted, does give the contractor some protection against such claims.

• Some statements of limitations go further by providing for inflation.

• If historical inflation data are any indicator, roofing costs over the next ten to twenty years can be expected to rise dramatically. By limiting its liability to the original costs, the manufacturer quite intentionally makes no provisions for inflation.

• With a limitation to the original cost, it is probable that the warranty will cover only a fraction of the replacement cost in the event of a total failure near the end of the warranty period.

• Again, you must understand that the warranty covers only the manufacturer's materials and the labor used to install them.

• It does not cover insulation, flashings, decking, or other items that might need to be replaced or repaired as a result of a failure.

fastfacts

➤ *Another statement that is generally part of warranties is that the cost of the repairs shall not exceed, in the aggregate, and over the life of the warranty, the sum of the original cost of the materials supplied and the labor used to install such materials.*

➤ *This has the intended effect of accumulating the costs of all repairs over the warranty period, and cutting off the spigot for this outflow of cash once the original dollar amount of the deal is reached.*

➤ *Without the in-the-aggregate clarification, the limitation could be interpreted to apply separately to each incident claim.*

- Consequently, a roofing contractor and an owner must be aware of these limitations in order to protect the contractor from spending profits defending unwarranted claims.

- At first, this situation may look like an additional bother for the roofing professional.

- However, it is preferable to take a proactive stance and turn the situation into a positive by combining warranties with maintenance.

Building Maintenance

- Many building owners are not careful and diligent about the maintenance of their roofs.

- A lack of maintenance can cause drastic problems for all of the parties who have been involved with a roof.

- The damages caused by poor maintenance often receive little attention in construction defect litigation.

- Maintenance can receive far too little attention in all phases of the construction and lifespan of a building.

fastfacts

➤ *Do not let the manufacturer's warranties become a problem—
sell your own warranties.*

➤ *The roofer can also provide warranties at various levels
that cover the job well beyond those of the manufacturer.
Warranties can be a healthy source of income.*

Encouraging Maintenance Programs

• Roof maintenance is one of the most neglected areas in the roofing industry.

• This is unfortunate, since maintenance can be one of the biggest money savers for building owners, and one of the most profitable areas for roofing contractors.

• Many times, when maintenance of roofs is contracted, it is not diligently performed.

• Poor roof maintenance is so prevalent that, in many cases, the only time the roof receives attention is when it leaks.

• Roofers often overlook this situation because they are absorbed in reroofing or the installation of new roofs.

• Perhaps the main reason for the void in roof maintenance is a general lack of awareness on the part of building owners, maintenance staff, and roofing contractors.

• Another deterrent is the roofing bond or guarantee by the roofing material's manufacturer and roofing contractor.

• Building owners often believe that these bonds or guarantees are all-inclusive and last forever, or at least 20 years.

• In the good old days, some did. Today, however, most guarantees are impressive on the surface, but in the fine print state that the roof must be kept watertight, or repaired by the roofing contractor if it leaks.

• Awareness of roof maintenance should begin the day the roof is completed and should become part of the overall building maintenance program.

- The standard two-year guarantee should be explained so that the building owner knows exactly what it does and does not cover.
- Begin roof maintenance with a visual inspection in early spring and late fall on a regular basis to determine if any damage to the roofing membrane or its components has occurred during the season.
- Roof maintenance is supposed to detect minor problems and arrange for their repair.
- Minor expenditures can eliminate the need for extensive repairs or premature total roof replacement.
- The cost of roof maintenance varies depending on the size, location, and design of the structure.
- Roofs that are well maintained from their completion through their lifespan have a lower maintenance cost than those that are permitted to deteriorate for several years before maintenance begins.

APPLYING MAINTENANCE COATINGS

- The roofing contractor can guide the owner away from the short-term, let-it-sit-there, throw-a-patch-at-it solution by pointing out the long-term savings that can be realized by including the application of a complete maintenance coating in the maintenance contract.

fastfacts

➤ *The price of roofing materials will continue to increase dramatically, and there is even concern that the availability of products will become an issue.*

➤ *If these two factors are combined, roof maintenance becomes very important to the building owner, and a very viable business for the roofing professional.*

➤ *Roofing contractors need to join other trades in the construction industry and develop guarantees that are explicitly defined and that detail the maintenance that is required by the owner and the consequences if that maintenance is not performed.*

- It is not always an easy sell, given the preference of some owners for the cheapest available option.

- This kind of straight talk, however, can build trust and eventually lead to increased sales and customer satisfaction.

- The straight-talk strategy also requires that the roofing contractor stay current on the technologies affecting business in today's market.

- The professional must keep abreast of the latest available information if she or he is to respond correctly to customers' needs.

- A decade ago, conventional systems were essentially the only choices on the market.

- Today that picture has changed radically, and new products are introduced almost constantly. Evidence of this fact can be found in the success of today's innovative synthetic systems, which offer performance comparable to that of traditional products and offer extra benefits, such as flexibility, increased durability, and even energy savings.

- There are a variety of different options in roof maintenance techniques.

- An objective listing of different options should include both conventional systems and the non-conventional systems presently available.

- Conventional choices include resaturants, asphaltic coatings, and aluminized asphalt coatings.

- A resaturant is described as a penetrating type of oil. It is recommended for improving the performance of weathered BUR.

- Available for half a century, resaturants are a quick, inexpensive way to fix an aging roof.

- Resaturants, which are composed of oils, are usually applied over roof membranes where the original oils have leaked out as a result of exposure to the sun.

- It is believed that these replacement oils revitalize the old roof.

- The shiny, black appearance of a freshly applied resaturant gives the customer immediate psychological satisfaction.

- The utility of resaturants, however, remains a controversial question.

- A study on resaturants conducted by the National Bureau of Standards demonstrated that the materials did not penetrate deeply enough to be effective, and that the mechanical performance of the resaturated roof did not improve enough to be worthwhile.

- Asphaltic coatings are the inexpensive traditional solution.

- They are composed of asphalt, either in a cutback solvent or emulsified in water.
- Asphalt membranes serve as a protective coating. However, asphalt is extremely susceptible to ultraviolet (UV) radiation and is not long lasting.
- Aluminized asphalt coatings are a more expensive solution.
- They maintain their popularity because of the reflectivity associated with them.
- These oil-based coatings impart a bright silver appearance.
- Usually within one year, however, the aluminum pigments oxidize, leaving a dull gray color.
- This causes the coating to lose its reflective capabilities, which decreases the energy-saving benefits.
- The asphalt in these particular coatings makes them especially prone to UV degradation.
- Today, non-conventional, liquid-applied maintenance coatings are gaining acceptance because of their practicality.
- Modified-bitumen roofing (MBR) sheets can be used over traditional BUR systems, as well as over the new single-ply systems, MBR sheets, and polyurethane foam.
- These non-conventional synthetic coatings place little extra weight on the roof deck or its supporting members.
- Even though they are initially more expensive than their conventional counterparts, these newer coatings are more effective in the long run, primarily with ethylene propylene diene terpolymer (EPDM), or aged BUR and MBR, because they eliminate the need for radical reroofing.
- Roofs can be recoated without adding significant weight to the roof load, and the coatings can be formulated in white to impart increased reflectivity to the roof.
- Besides contributing to longer service life by reflecting UV rays, the white color keeps the roof surface cooler, which reduces the demand for air conditioning inside the building.
- The advantages of longer service life and decreased energy costs are the driving forces behind the popularity of reflective roof coatings.
- The most distinguishing feature of liquid-applied roof coatings is that they are elastomeric, or flexible.
- This flexibility enables the coating to withstand substrate movement caused by severe temperature changes.

- It should be noted that thermal shock is not limited to northern climates.
- A sudden thunderstorm in Florida can quickly reduce the roof surface temperature by as much as 100°F.

Non-Conventional Roof Coatings

- There are three types of non-conventional roof coatings: solvent, water-based, and 100 percent solids.
- Solvent coatings use organic solvents and can be applied over a wide range of temperatures. They dry quickly when humidity is high, but they usually contain large quantities of flammable, and often highly toxic, organic solvents.
- Extra care must be taken when transporting, storing, and applying these coatings. In addition, solvent-based coatings are associated with staining problems on certain asphalt roofs.
- Water-based coatings use water, rather than solvent, as the carrier.
- This eliminates the flammability and potential toxicity hazards linked to solvent-based systems.
- Cleanup is done inexpensively with soap and water.
- These coatings can even be applied to a damp substrate without losing significant performance.
- The main limitation of water-based coatings is that application is affected by weather conditions.
- They should not be applied when the temperature is below 40°F, or when rain is imminent.
- Since water based coatings contain no organic solvents, they are less likely to create staining problems on aged roofs.
- The 100 percent solids coatings have virtually no solvents in them, but they must be applied with special equipment.
- In addition, they are sensitive to changes in temperatures.
- Their capacity to form a strong protective membrane depends on the maintenance of the proper reaction temperature and the mixing ratios of the components.
- If either is wrong, this type of coating cannot dry properly and may even gel prematurely, which could cause a defective membrane.
- Within these three classes of liquid-applied coatings, there are four kinds of products available.

- Each is based on a different chemistry: urethane, Hypalon, co-polymer acetate, or polyvinyl acetate, and acrylic.
- These coatings are not asphaltic, and thus they are quite different from bituminous emulsions or resaturants.
- The urethanes are usually supplied as solvent solutions and can be either one-component or two-component systems.
- The advantage of two-component coatings is that they offer quicker curing than one-component coatings.
- A good deal of skill is required to obtain the proper ratio of the two components.
- Urethane coatings are also rather expensive compared to other classes of coatings.
- Hypalon coatings are based on chlorosulfonated polyethylene.
- While normally supplied at low solids to permit application with a spray gun, these coatings are comparatively expensive and offer limited durability.
- Polyvinyl acetate copolymer-based coatings are less expensive, water-based systems.
- They can be used when a white reflective surface over an existing roof is desired and long-term tolerance of thermal shock is needed.
- These coatings are not recommended for many uses because they offer comparatively poor durability. Consult with the architect and manufacturer.
- In addition, polyvinyl acetate copolymers tend to hydrolyze, which also affects durability.
- Acrylic roof coatings can be supplied as either water-based or solvent-based products.
- They provide an excellent combination of high performance and durability for a relatively modest cost.
- Acrylic coatings provide superior resistance to UV radiation, as well as good resistance to dirt pickup and water ponding.
- An acrylic coating forms a seamless, breathable membrane.
- It can be applied readily to irregularly shaped roofs.
- For best results, acrylic coatings should not be applied when there is high humidity, or if rain is imminent.
- Almost all roof maintenance coatings are best suited for one type of project or another.
- Some are recommended only for specific situations.

- If there are doubts about the best overall option, require that the owner retain an architect.

- The benefits of being able to point out the performance and cost features of the many available products cannot be stressed enough.

SELLING WARRANTY AND MAINTENANCE CONTRACTS

- You have a limited duty to warrant the products you provide to the public as a roofing professional; that is a given.

- However, you also have the opportunity to provide a warranty service for your clients that offers them protection well beyond the obligatory, statutory requirements.

- You will have to think this through, determine what your firm can offer, and what your particular clientele needs.

- The idea is that you must provide a limited warranty; however, your clients can purchase a more extensive warranty if they prefer.

- For example, the warranty might cover failure of plumbing vents from the top surface of the roof sheathing up, breakage of tile (with limitation of visits), and all other elements of the roof.

- This may seem a bit risky at first, but the package should be available only to those customers who choose to sign up for your company's maintenance service contract.

- Contracts have to be designed specifically to fit your company's mix of products and clients, and also to fit your staff or the staff you intend to hire.

- This is a proactive approach that benefits all parties.

- The average building owner is not aware of how much punishment a roof takes. Water, sun, wind, foot traffic, trees: most people are not aware of just how tough a roof has to be, or that it must be cared for in order to extend its lifetime.

- In the case of commercial buildings, the owner should be made aware of how difficult it is for professional building maintenance firms to hire people who are both knowledgeable and diligent when it comes to roofs.

- The basic proactive approach is that your customers may purchase an extended warranty from your company, but only if they sign up for the maintenance service.

fastfacts

➤ *Although it may take some changes in the roofing professional's approach to everyday business, selling maintenance
and warranties can create excellent customer relations and
lock in some strong profits.*

➤ *The basic idea is that your firm will visit the building twice a
year and perform a visual inspection from the ground, as well
from the roof.*

➤ *The inspector will provide the client with a signed inspection
sheet and an estimate for service and repairs.*

➤ *The service and repairs may include treating galvanized flashing for rust, replacing damaged gutters, drains, and a general
cleaning of the gutters, drains, and scuppers.*

➤ *Use photos of damage to get the client's attention; nobody
wants to throw money away, and how many building owners
ever get on top of their buildings?*

- This closes two sales, and the sale of any future work the roof may
need before it requires replacement.
- The arrangement is excellent; you have developed an ancillary
cash flow, and your customers have long lasting products.

CHANGE ORDERS AS A PROFIT CENTER

- Change orders are another potential problem area for roofers during any construction project.
- In theory, design drawings and specifications are executed by the
design team for the building owner.
- These plans and specifications are the blueprint for what is to be
built. Each party submits a cost for its component and the building project is started.

- However, anyone who has any building experience knows that the original plans and specifications are almost always changed during a construction project because of additions to the structure, alterations in the design of the building by the designer and owner, or onsite requirements for changes in the design because of unforeseen construction situations.

- In a perfect world, the construction of the building would be orchestrated by an onsite project manager who is extremely diligent, with good weather and good performance by all other parties, so that there are no surprises for the roofing contractor when he or she arrives.

- An example of a surprise that the roofer might find is an alteration in framing, such as changing the slope direction, which causes problems with drain locations.

- The plumber and the general contractor's project manager have not seen this, but the roofer's project manager spots it on a pre-installation site visit.

- The roofer's project manager prepares a change order, rather than arguing with the general contractor or the design team.

- The principals accept the change order.

- The roofer has brought the job back on track, prevented future infiltration of water and possible construction defect litigation, and added to net profit in a few moments time.

- In most cases, the communication is adequate in new construction jobs, and relatively few problems occur. However, the roofer cannot count on clear communications being automatic.

- Yet, with all of this coordination and communication, roofers are often the last to know about many things. Since the drain is the plumber's responsibility, the roofer's input is rarely sought.

- Even if everyone is consulted and all perform their expected duties, after the plumber completes his or her tasks on the roof and the roofer reassembles the drain to the existing roof system, who inspects the final drain assembly?

- Because the plumber's responsibility is almost nonexistent on the roof, he or she is naturally reluctant to go back to the job to check the roofer's work.

- Very often, plumbers do not have any real expertise in roof drains, so they probably could not find any problems even if they undertook an inspection at completion.

- Thus, defects are often ignored until there is a leak. Precautionary measures can be taken if proper preparation is observed beforehand.
- The important thing is to make this a part of your ongoing bidding process.
- Review bid specifications carefully, and then submit change orders to the principal and the design team to apply for addenda aimed at alteration or replacement of faulty design criteria.
- A page of change orders can add a good deal of money to the bottom line of any estimate, and this is often money that is earned at terrific profit margins.
- Pay close attention to developing the sale of change orders as an ordinary part of doing business, and always get a valid signature to back up the request.

MAKING MONEY AS A ROOFING CONSULTANT

Within the last decade or so, construction defect (CD) litigation has become a huge business on the periphery of the AEC (architecture, engineering, contracting) industry. Many of you are well aware that this is an immense web of complex civil actions, and the cash flow is immense. If your firm has been sued directly or in a cross complaint, you are well aware that CD litigation can be a long, drawn-out process.

- Taking a proactive approach to roofing and selling through maintenance contracts will help to keep your firm from being ensnared in the complex web of litigation.
- The ideas in this book must be implemented in a professional manner, and continual diligence must be applied to make them a seamless part of your company.
- Since they are extremely vulnerable to the elements, roofs have been a large part of the tidal wave of CD litigation.
- Typically in these lawsuits, the plaintiff has noticed something wrong with the building, generally water intrusion.
- The owner hires a plaintiff's lawyer. The plaintiff's lawyer hires an expert team, in which one member is typically a roofing expert.
- The plaintiffs sue the developer or general contractor, who hires lawyers, who then hire more experts.

- The developer sues the subcontractors in cross complaint, and their lawyers hire more experts and file notice with the insurance carriers, who may hire more lawyers and experts.
- This is a very brief rundown of a possible situation. Naturally, the circumstances can vary greatly, but it is obvious that CD lawsuits require many hours from roofing experts.
- The discovery on the roofs is one of the parts of the fieldwork that is most heavily laden with billable hours.
- The time billing for this work is typically set at a fair rate for the roofing expert, and there can be a good deal of it available.
- CD work is only one part of being a roofing consultant; there are numerous other services that your company can provide:
  - Inspections for realtors and corporations.
  - Feedback for architectural firms.
  - Expert witness at trial.
  - There are many ways to offer your knowledge.

How Consulting Works

- Typically, you will be hired to investigate the condition of an existing roof.
- Generally speaking, there are two ways in which this is done:
  - Visual inspections from the ground and on the roof.

fastfacts

➤ *These parallel cash flows are a part of your general business plan; at the same time, you will reap excellent side benefits from this work.*

➤ *A big asset, for example, your understanding of what your roofing firm should not do, will increase dramatically.*

➤ *Also, you will have the opportunity to associate with numerous people to whom you can market all of your other services.*

- Thorough investigations during which parts of the roof area are opened up.
- These are often referred to as nondestructive or nonintrusive, and destructive or intrusive site inspections or investigations.
- As with everything related to the AEC industry, different people use different terminology.
- The usual scenario for discovery on site is that the plaintiff experts walk the site, and produce a rough list of possible construction defects.
- This is usually called the preliminary defect list.
- Destructive work is not typically a part of this process.
- The defense team may then walk the site for a round of nondestructive testing, and then the two parties will meet to discuss their findings.
- At that time, they often agree on what parts of the buildings need to be cut open in order to determine what has actually caused the listed defects, which cannot be learned from nondestructive observation.
- The next step in the process will be to schedule the first round of destructive testing.
- During this time, parts of the roof may be opened up in order to determine what has caused the leaks and other problems that have been observed.
- In certain cases, instruments are used to avoid the costly process of cutting buildings open, and then patching them back up, involved in destructive work.

Performing Nondestructive Testing

- Millions of dollars are spent each year for roof repairs caused by leaks.
- This creates an ever-increasing demand for more accurate moisture damage assessment.
- Roof leaks that result in wet insulation, yet cannot be seen as a drip, cause real problems for roofers.
- A successful moisture analysis can reduce overall repair costs significantly, while developing and securing mutually beneficial relationships between roofing professionals and their customers.

- Nondestructive testing (NDT) is designed to locate moisture in roofing systems, including interply and substrate moisture, without causing physical damage to the roof's surface or substrate.
- Wet insulation in roofing systems is often compared to a cancer.
- NDT acts as an x-ray or biopsy of the roof to determine the location of the moisture.
- If left untreated, this moisture causes the roof membrane, fastener system, and structural deck to deteriorate.
- Figure 4.1 shows some elements of nondestructive testing. Descriptions of the most popular NDT methods follow.

Infrared Testing

- Infrared instruments, both hand-held and aircraft-mounted, can be used to detect trapped moisture by identifying areas of heat loss through wet insulation.
- One of the advantages of infrared testing is that large areas of the roof can be scanned relatively quickly.
- Using an infrared-scanning camera, the operator takes a photograph, called a thermograph, that indicates energy loss due to wet insulation.
- However, it is important to take into consideration the following disadvantages of infrared testing:
  - In most cases, tests must be conducted at night.
  - Exacting weather requirements dictate when tests can be run to achieve accurate results.
  - The cost of the equipment is significant.
  - Readings can be affected by hot air escaping on a roof, heat sources suspended below the roof, differences in insulation thickness and BUR, wet surfaces, and ponded water.

Nuclear Testing

- Moisture is made up of hydrogen and oxygen.
- Nuclear moisture detection operates on the principle of neutron moderation, which means that fast neutrons emitted by a radioactive source are slowed by hydrogen in the roof.

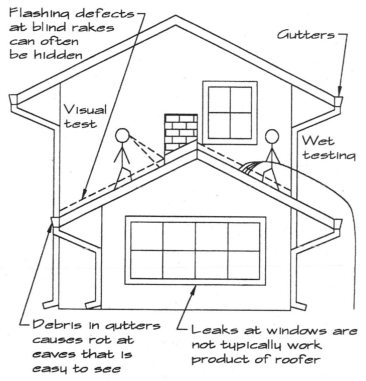

FIGURE 4.1 Nondestructive testing.

- The slowed neutrons are detected and the count displayed is proportional to the hydrogen content.
- Nuclear testing also has its disadvantages:
  - Since asphalt and bituminous roofing compounds contain hydrogen, variations in material density can be misinterpreted as moisture.
  - Nuclear surveys are very slow.

- Nuclear testing is regulated by the Nuclear Regulatory Commission (NRC), which requires operators to be licensed, and strictly controls the use, transportation, storage, and disposal of these instruments.

Impedance-Method Testing

- The impedance method combines the principles of resistance and capacitance.
- These instruments operate by emitting low-frequency electronic signals from rubber electrodes fitted at the base of the instrument.
- When there is no moisture present in the substrate, these signals are insulated from each other and no reading is recorded.
- When moisture is present, the electrical conductance is dramatically increased and the circuit is completed, giving instant readings.
- The greater the moisture content, the higher the reading.
- Advantages include:
  - Versatility and ease of operation.
  - Instant and continuous readings.
  - Ability to trace the path of leaks and pinpoint the place of entry.
  - No effect from temperature differentials caused by wind, vents, or variable roof layer and asphalt thickness.
  - Safe and easy to transport and use.
- A disadvantage is that these meters do not operate on roofs with a metal cap sheet, through ponded water, or on inverted roofing systems without the removal of ballast and insulation.

PROVIDING PREVENTIVE MAINTENANCE

- Consulting in CD litigation is not the only area where non-intrusive work can be a valuable service.
- The lifespan and energy-saving performance of a roofing system are greatly influenced by the presence or absence of a roof maintenance program.
- For this reason, non-destructive testing is becoming increasingly popular among roofing professionals.

- These cost-effective diagnostic methods provide vital information that can expand business opportunities and ensure long-term profit potential.
- Many roofers agree that more than one method of study is necessary to properly survey for roof leaks.
- Each system complements the others, and can be used for verification of their findings.
- Anything that does not show up using one method usually shows up when another method is used.
- However, there are times when water testing and intrusive work must be resorted to in order to fully understand a roof problem.
- Destructive testing is fairly straightforward for a veteran roofer (Figure 4.2). The problem area is studied before the intrusive work is done, and the experts instruct the mechanics doing the work what areas to open and in what order.
- The experts observe this with care, photographing, taking notes, and sometimes running water, in order to understand the exact problem, how it was caused, and how it can be fixed in a cost-efficient manner.
- There is no substitute for expertise and roofing knowledge.
- No machine can replace the roofing contractor's experience and knowledge about the many different roofing systems.
- For example, some roofs require insulation that is not butted firmly.
- A roofer must have such knowledge when inspecting the roof, so that false readings do not send the team on a wild goose chase.
- There is a host of knowledge that only years of hands-on experience can bring to the table, and that is why you should be well paid for using this base of information to help building owners work through their difficulties with the various concerned parties.

PROMOTING ROOF ACCESSORIES

- There are several roof accessories that roofing contractors can install as part of their normal work operations to help their profit picture.
- For example, the installation of a roof walkway between equipment located on the roof can prevent damage from foot traffic (Figure 4.3).

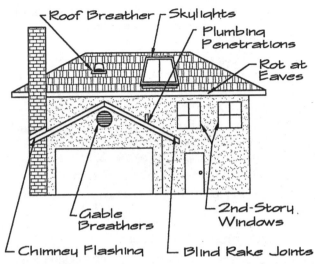

- → Skylights are typically well-flashed, but removal of tile at curb flash is sometimes needed.
- → Breathers in walls are not typically the roofers' work product.
- → Roof breathers will leak if lapping is not well executed.
- → Eaves and open rakes must be studied closely to allocate resultant damage – – home owner and other trades can contribute to cause of damage.
- → Stains below 2nd-story windows are not typically resultant of roofers' work product. Careful wet testing with destructive exploration is often required.

FIGURE 4.2 Destructive testing.

FIGURE 4.3   Roof accessories/walkways.

- Roof walkways, especially when combined with handrails, also provide surer footing for future visitors to the rooftop, which can be a valuable safety feature for the building owner.
- Roof curbs enable roofing contractors to effectively flash a roof penetration.
- Some curbs that are installed on a metal roof panel can have a water diverter and matching ribs welded into the curb base.
- Other roof accessories that enhance safety and protect a building owner's investment are access hatches, smoke hatches, and gravity ventilators.
- Access hatches allow workers to get onto the roof safely to install or repair rooftop equipment.
- Several states have upgraded their fire code requirements, and architects are specifying that smoke hatches be installed on an increasing number of projects.
- Smoke hatches typically have a release mechanism that is attached to a fusible link.
- When the temperature inside the building reaches a predetermined level, the fusible link breaks.
- This activates the release mechanism and the hatch opens to release the smoke.
- Releasing the smoke helps firefighters bring a fire under control quickly, safely, and with less water damage to the building and its contents.

- Gravity ventilators provide for the safety and comfort of building occupants by regulating the temperature and air quality. They also help protect the building's contents from moisture damage.

- When properly utilized in conjunction with adequate air-intake devices, gravity ventilators can provide an inexpensive way to ventilate a building.

KEEPING SNOW FROM SLIDING

- Snow-retention devices can be invaluable when snow must be prevented from sliding off roofs.

- Snow guards come in many types, but they are all too often omitted from new construction projects in order to save money, and dismissed on existing structures on the grounds that the event that prompted their consideration in the first place might not happen again.

- On textured roofs or structures with limited liability exposure, these justifications might be reasonable, but on metal and slate roofs, this kind of thinking can be dangerous as well as costly.

- While snow initially accumulates on even the steepest of metal roofs, it is not going to remain there. Unless it is restrained and allowed to dissipate as snowmelt, or the amount of snow that can slide free at any one time is restricted, the accumulated snow can cause considerable injury or damage.

- Even low-slope metal roofs experience snow movement. Although research indicates that sliding is unlikely until the roof slope is 14 degrees or more, snow massed on shallow-pitched roofs can behave like a glacier and creep slowly down slope, even on a very gradual slope.

- The first line of defense should be good planning during the design of a structure.

- This can reduce the potential for harm to persons, property, and even the building itself when snow or icing is anticipated.

- A host of criteria determine the placement of snow guards on a metal roof.

- Climate, obviously, is the first of these because it affects the amount, frequency, and physical characteristics of the snowfall, e.g., wet, dry, or mixed with sleet or ice.

- Roof height, shape, type, pitch, lengths of runs, pan widths, locations of other physical features such as gables, changes in roof elevation and pitch, vents, stacks, chimneys, other penetrations, the directions of prevailing winds, and even the presence of snow-making machinery at ski resorts all have an impact on snow-retention measures.

- The purpose for which snow or ice is to be retained also must be evaluated:

  - Are building occupants, patrons, or passersby to be protected?

  - What about foundation plantings, vehicles, adjacent structures, or equipment?

  - Is there a need to control snow buildup on roofs below, or to protect gutters or stacks?

- Of equal importance to many owners, and every architect is the visual impact that these devices have on the structure.

- Given all of these variables, one can see that device placement certainly requires some thought.

- The traditional fastening method, used for almost 100 years, requires a pair of prongs to be driven through the seam.

- This method often results in deteriorated metal, sheathing, and framing.

- While this type of snow guard is still used on slate roofs, the three most popular designs for metal roofs are fences, surface-mounted devices, and seam-mounted devices.

Fences

- Fences are field-assembled systems designed to hold snow at the eaves.

- They consist of rods and various other components.

- The height of the bottom rod above the roof surface is such that these systems are not particularly effective in retaining ice or light accumulations of snow.

- Care must be exercised during design.

- Unless sufficient rows are employed, fences tend to concentrate snow loads rather than distribute them over the roof.

Surface-Mounted Devices

- Surface-mounted devices are secured to the pan of the metal roofing system with solder, fasteners, and adhesive sealants.

- While effective in retaining both ice and snow, many of the devices are not well engineered and lack material strength (Figure 4.4).

- Quality devices are particularly useful when there are no suitable seams to which a seam-mounted device can be fastened.

- Unfortunately, the use of adhesives is limited by weather conditions. In hot weather, temporary supports or tape may need to be employed until the sealant sets up.

- In cold weather, the adhesive may not cure properly.

- Generally, a minimum temperature of 50°F for a period of 30 days is required for a professional application. In any case, the surfaces must be clean and dry.

Seam-Mounted Devices

- Seam-mounted, one-piece snow guards are secured to the standing seams.

- This one-piece device is clamped to the metal roof seam with stainless steel set screws and installed with a ratchet and hex socket.

- This makes the device simple to install and easy to retrofit on existing roofs.

CONTROLLING BIRDS

- On some roofs, bird control is serious business. Gimmicks, toys, or other gadgets seldom work.

- If your client has this problem, you might suggest the installation of a stainless, humane bird barrier, such as the type shown in Figure 4.5.

- As you begin to notice services that you can provide for your clients that will enhance your cash flow, more of them will fall into place.

- Take one step at a time, but do not procrastinate. Remember your goals and pursue them diligently; this will add a whole new sense of interest and vitality to your business.

✓	Item
	Review the structural capacity of the roof to verify that it is adequate to support the loads that might develop.
	Place snow guards at or within the interior face of insulated exterior walls to reduce the potential for ice dams at the eaves.
	Space snow guards with the goal of holding snow in place, rather than trying to catch it at the eaves. On long runs or steep pitches, this means spacing the devices at suitable intervals up the slope. It is easier to retain snow or ice on a roof than it is to resist the dynamic loads generated once it breaks free.

FIGURE 4.4 Snow load guidelines. Use this checklist for defining your work and leave plenty of room for notes.

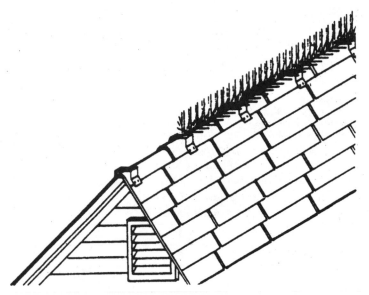

FIGURE 4.5 Stainless humane bird barrier.

THE FUTURE OF ROOFING

- The traditional roofing practices described in this book are by no means on the way out. One-ply, urethane, fiberglass, tile, slate, or conventional commercial BUR systems, will be around for a long time.

- Those that remain have done so because they evolved as roofing systems that meet the needs of commercial and residential roofing contractors. In this ever-changing world of construction and architecture, however, new concepts, systems, materials, products, and ideas are definitely in use with more on on the way.

- An excellent example of a brilliant and far-reaching new concept is fabric roofing.

- In the end, being able to protect the tops of our built structures from the elements with lightweight sheeting is a powerful concept, a step forward in the evolution of one-ply systems.

- Permanent fabric roof structures are past the experimental stage.

- Fabric roofing is undergoing the technological evolution that takes place in every segment of this industry.

- They are not the best solution to every design program, but they are an excellent option when the design requires long, clear spans.

- For example, the Denver International Airport is a unique architectural statement.

- The distinctive silhouette above the terminal roof, which shows 17 sets of pure white peaks, evokes a sense of the Rocky Mountains (Figure 4.6).

- When a clear span exceeds 100 to 150 feet, fabric structures generally cost less than traditional structures. Proponents of fabric structures cite the fact that the weight and cost of the supporting structure, per unit of plan area, do not increase with the roof's clear span, as in conventional construction.

- Furthermore, the cost advantage occurs because the fabric envelope performs the function of several conventional building components, including roofing materials, roof deck, insulation, structure, wall cladding, waterproofing, acoustical absorber, interior finish, and daytime light source.

FIGURE 4.6 Denver International Airport evokes a sense of the Rocky Mountains with its fabric roof.

- It should be noted that fabric structures usually require specialized engineering, which can make them a more economical option for larger structures than for smaller structures.

- Special consideration also should be given to the use of fabric structures in cold climates, despite the fact that additional insulating materials may need to be used to expand their climatic range.

- Fabric structures must be held in tension to be structurally stable.

- There are two methods to accomplish this goal. Internal air pressure can be supplied resulting in an air-supported structure, or the fabric can be mechanically pretensioned from a mast or frame to create a tension structure.

- Curves are a distinctive feature of fabric architecture because, in both air-supported and tension structures, they are necessary to translate applied loads into tension forces and to ensure the stability of the structure.

- Tension structures use a double curvature that is anticlastic, or in opposite directions.

- An air-supported structure also has a double curvature, but the curves are synclastic, or in the same direction.

- Air-supported structures are engineered with an increased interior pressure of 0.3 percent to inflate and post tension the fabric roof.

- A network of cables across the roof relieves stress in the canopy fabric by transferring wind lift and inflation loads to the ground or walls.

- A compression ring on the ground or walls resists the anchor loads of the cables.

- Doors are airtight and small fans operate continuously to maintain interior air pressure.

- Practically speaking, there are no limits to the area an air-supported structure can enclose.

- The cost per square foot of conventional construction methods increases dramatically as clear-span distance increases.

- However, with air-supported structures, the cost per square foot actually might decrease.

- Fabric in a tension structure is pretensioned into anticlastic shapes.

- Structural stability results from the double curvature.

- Applied loads from any direction are resisted by a modified stress field in the fabric canopy.

- Tension structures are supported by masts, arches, cables, or edge beams.

- They create a dynamic visual appearance because they allow endless variations of shape and form.

- Unlike air-supported structures, tension structures can be open to the outside because internal air pressure is not required for stability.

- Fabric roofing is not the only material that the roofing contractor will consider in the future.

- With computers as design and engineering tools, products can be invented and tested more quickly than at any time in history.

- New products and roofing structures are now in the design and experimental stage. They will be the cutting edge in the field soon enough.

- Photovoltaic roofs, which have the ability to produce electricity from the sun, are already available and will become important products as the population doubles and the cost of energy keeps going up.

- To keep abreast of the changes in the roofing industry, the roofing contractor should read such publications as *Professional Roofing, Western Roofing, Insulation and Siding,* and *Roofer Magazine.*

- In addition, as part of becoming more professional in your business, you might also benefit from joining organizations such as the National Roofing Contractors Association.

- This is a very dynamic time in the history of the roofing industry.

- There are many directions for growth available to the roofing professional. A healthy mix of old tried-and-true products and services, and work in the forefront technologies like fabrics and photovoltaics may be the best route for your firm.

- The new products and services might just inspire everyone at the firm, as learning keeps us all vital and interested.

- The door to success is wide open, but the decisions are all yours to make.

5

BUILT-UP ROOFING

Built-up roofing (BUR) uses bitumen asphalt or coal tar as the waterproofing and adhesive agent that works with the felts to create a membrane that sheds water off the building. Combined properly with flashing and any adjacent, vertical surfaces like stucco or wood siding, built-up roofs create a long-lasting, fair-priced product for the customer.

BUR is the oldest of the modern low-slope roofing systems and is still a very common application. It remains widespread partly because of tradition and partly because if the layers are applied correctly, it provides good protection at a fair price.

- Figure 5.1 illustrates how built-up roofs are put together.

- Typically, they consist of a protective top surface, the bitumen coats layered with reinforcing felts, and the bottom layers of insulation materials.

- Asphalt and coal tar are the bitumens of choice.

- These thermoplastics become more fluid with heat and revert to a more solid state as they cool.

- Built-up roofs are an excellent choice for many roofing applications.

- As in all of the installation sections about the various types of roofs, built-up roofing starts with general information and continues through maintenance.

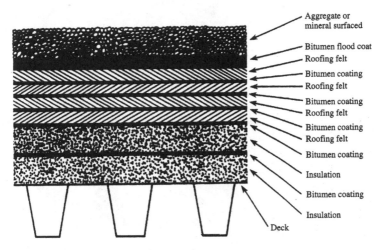

Aggregate or mineral surfaced

Bitumen flood coat

Roofing felt

Bitumen coating

Roofing felt

Bitumen coating

Roofing felt

Bitumen coating

Roofing felt

Bitumen coating

Insulation

Bitumen coating

Insulation

Deck

FIGURE 5.1 Layers of a built-up roof.

GENERAL INFORMATION

Materials

- Built-up roofs can be very satisfactory roofing systems with a good lifespan to cost ratio and simple maintenance requirements.
- Since built-up roofs are common and relatively simple systems, do not make the mistake of thinking that BURs can be applied without being diligent and professional.
- BUR systems have definite installation requirements that start with materials and materials handling.

The following guidelines for the handling, protection, and storage of all materials should become a regular part of company activities and be strictly followed at all times.

Materials Procedures at the Job Site

- Always protect materials from the elements, vandalism, and theft.
- Store all cartons, insulation, drums or cartons of asphalt, and cans of cement on raised, level platforms.

- Indoor, locked storage, such as a built structure or shipping container, is the preferred method for stockpiling materials.
- If no structures are available or open structures are used, protect materials with waterproof tarpaulins or other rugged safeguard devices.
- The sun can have a strong reaction with materials.
- Improper storage of packaged asphalt can result in spills and contamination.
- Store all roofing and flashing materials in a dry place.
- Moisture in the asphalt from improper storage causes foaming when heated in the kettle, which can result in injury to the roofing mechanic.
- Store roll goods on end and on a clean, raised platform to keep the ends of the rolls free from foreign matter.
- Plastic covers, stretch-wrap, and shrink-wrap are not intended for job storage, since moisture can condense within and on the material inside.
- Protect materials from the elements with waterproof tarpaulins.
- Store solvents and solvent-containing material in a cool, dry, fire-safe area.
- Unload and handle all roofing material with care.
- Dropping rolls, roof insulation, cans of adhesives, and roof accessories can damage these components enough to cause unsatisfactory application and performance.
- Note any damage on the trucking firm's bill of lading.

fastfacts

➤ *BURS will give the customer excellent results if you sell a maintenance contract with the new or re-roofing product so you can make certain that it is well cared for.*

➤ *BUR materials must be safe from damage, and they must be kept dry. The use of wet or damaged materials can contribute to the massive failure of BUR systems.*

- Notify manufacturers immediately if materials are not received in satisfactory condition.
- Do not heat any supplied BUR materials with an open flame.
- Keep the temperature of all roll goods above 40°F for 24 hours prior to application.
- MBR products are considerably easier to install when maintained at temperatures above 50°F.
- At the end of the workday, cover all roofing materials remaining on the deck.
- Observe all fire precautions involving the storage and handling of roofing materials.
- The importance of taking care of materials cannot be over emphasized. They are expensive, and missing or defective materials cause slowdowns in production, callbacks, construction fixes, and can even trigger construction defect lawsuits.
- An excellent strategy for improving the timely arrival of roofing products at the jobsite and the protection of them after they are checked in, is to give one person, who will be at the jobsite throughout the project, responsibility for all roofing materials.
- The BUR materials jobsite checklist (Figure 5.2) can be used by the foreman directly from this book.

BUR Equipment

A waterproof roof depends on heating the bitumen correctly so it can hold the BUR system together.

- There is a wide range of kettles offered today. You can search them out on the Internet
- Take care when choosing your kettles. Always maintain them and replace them when the time comes. Safe, reliable kettles can make the difference between profits and losses.

Equipment Guidelines

- Make the manufacturer's attention to safety a big part of your purchasing decision.
- Before transporting trailered equipment, verify that your vehicle can handle the load safely.

BUR MATERIALS JOBSITE CHECKLIST		
✔	ITEM	NOTES
	Cartons, insulation, asphalt, cans--on raised, level platforms	
	All tarps in place	
	Roll goods on end and on a clean, raised platform	
	Flashing materials in dry place	
	Plastic covers, stretch-wrap, and shrink-wrap removed	
	All materials protected from the elements	
	Packaged asphalt cans all tight, level, and safe	
	Solvents and solvent-containing material in a cool, dry, fire-safe area	
	All materials damaged by handling have been replaced	
	All manufacturers notified about materials not received in satisfactory condition	
	All damages found during receiving have been noted on the trucking firm's bill of lading or called and confirmed in writing	
	Temperature of all roll goods is above 40°F for twenty-four hours prior to application	
	All materials remaining on the roof at the end of the work day are protected	
	No open flames allowed for heating BUR materials	

FIGURE 5.2 Jobsite checklist for built-up roofing.

- Before transporting, check the kettle's suspension, connections, and safety lights.
- The kettle burner should not require full power to melt any dried material.
- It should take only a few minutes of low temperatures to create a vent around the heat riser.
- Remove all equipment and materials from the designated kettle area to keep them from absorbing the extreme heat.

- The attendant should keep a constant watch, but should never linger around the kettle.
- The combined heat from the kettle and the sun can cause heat-stroke or heat cramps in a short time.
- The best operator is one who has worked with the burner pump before and knows its behavior and peculiar habits.
- Make sure that everyone working around the hot stuff has properly protected their hands, face, and other body areas at all times.
- Remove all trash and debris from the kettle area.
- Do not allow heavy bitumen layers to build up on the kettle.
- Check continually for fuel leaks.
- In case of a kettle fire, close the lid and shut off the fuel supply.
- Keep a working dry-chemical extinguisher on hand at all times.
- Clean kettles save time and fuel.
- Clean kettles mean less sludge and produce a better quality of asphalt.
- Regular cleaning increases the service life of the asphalt kettle.
- A clean kettle is a safer kettle; trapped oxygen and light, volatile oils create dangerous fires.
- When using tankers, keep the ventilation system free and clear at all times.
- Follow the manufacturer's cleaning instructions for your kettles and tankers on a regular, calendered basis.

Kettle safety is so important that we have included this section under built-up roofs as a supplement to Chapter 3. The following are safety precautions and practices that must be followed when working with bitumen and a kettle (see also Figure 5.3).

BUR SAFETY

Kettle Safety Guidelines

- Place the kettle so that the lid opens away from the building.
- Place the kettle so that fumes are not blown into air ducts.
- Light burners by placing them on a noncombustible surface and pointing them away from equipment and materials.

FIGURE 5.3 Large pump kettle.

- Start the kettle at low heat and keep the flame straight down the tubes.
- Keep the kettle tubes covered with bitumen.
- Keep an ABC-rated fire extinguisher nearby.
- Keep the kettle clean by dipping it.
- Inspect hoses, fittings, and regulators daily.
- Shut off burners before refueling the kettle engine.
- Secure propane cylinders in an upright position at a minimum of 15 feet from the kettle.
- Wear certified protective equipment while working with hot bitumen.
- Avoid direct contact with asphalt and pitch fumes.
- Keep hot bitumen carrying or storage equipment free of any water or ice.
- Always rope off the kettle areas from intruders.
- No walking in hot stuff to avoid slips, falls, and burns.

- Daily end of shift procedure:
  - Turn over mop tubs and mini-moppers.
  - Close lugger lids to prevent the accumulation of ice and water.
  - Remove used mops from the roof, spin them out, and then place them on a non-combustible surface.
  - Lock up kettle burners or remove them from the jobsite.
  - Lock the kettle lid and outlet valve.
  - Protect the hot-stuff line that extends from the kettle to the rooftop with either a pipe covering or galvanized stovepipe.
- Do not overheat the asphalt.
- Make sure that the rooftop transport equipment, including hot luggers and mop buckets, has adequate insulation.
- Make certain that the material is hot enough at the point of application to permit good adhesion.
- The importance of job safety for roofers is a never-ending subject and it covers a great many areas of concern. From items as simple as wearing protective clothing to roping off the deck to prevent falls, the roofing contractor must always be diligent.
- A safety issue that relates to cold weather is the wind chill factor. We have included the following table to use for the protection of your crews (Table 5.1).

Cold Weather Guidelines

- During cold weather, it is absolutely essential that the entire roofing operation be completed as you progress each day.
- Do not phase cold weather applications except on temporary roofs.
- Check continually for good ply adhesion.
- Holidays or fishmouths do not have an opportunity to reseal themselves during cold weather.
- In cold weather work, it is absolutely mandatory that materials be delivered to the jobsite and installed in dry condition.
- Do not place materials in small piles on the deck—snow removal becomes difficult and the piles create snow fences.
- Store material in a heated warehouse or closed trailer just prior to installation.

TABLE 5.1 Wind Chill Factor

Wind velocity (mph)	Bulb ambient temperature (°F)								
	50	41	32	23	14	5	−4	−13	−22
	Equivalent temperature (°F) in cooling power on exposed flesh								
Calm	50	41	32	23	14	5	−4	−13	−22
5	48	39	28	19	10	1	−9	−18	−27
10	41	30	18	7	−4	−15	−26	−36	−49
15	36	23	12	0	−13	−26	−36	−49	−62
20	32	19	7	−6	−18	−31	−44	−58	−71
25	30	16	3	−9	−24	−36	−51	−63	−78
30	28	14	1	−13	−27	−40	−54	−69	−81
35	27	12	0	−15	−29	−44	−58	−71	−85
40	27	12	−2	−17	−31	−45	−60	−74	−89
45	25	10	−4	−18	−33	−47	−60	−76	−89
50	25	10	−4	−18	−33	−47	−62	−76	−90

Little danger	Increasing danger	Great danger

Danger from freezing of exposed flesh (for properly clothed persons)

- Yard storage tanks make it possible to mobilize at the last minute in bad weather.
- Remember that base sheets and other roofing materials are more brittle and less flexible in cold weather.
- Use winter-grade materials in cold weather applications.
- The completed roof is brittle in cold weather; keep mechanics and equipment off of the finished areas.
- Drain the kettle at the end of the day to prevent extra heating the next morning.

fastfacts

➤ *When roofing over concrete decks in cold weather, it is diffi-cult to wait until they cure.*

➤ *It helps to prime the deck with asphalt primer immediately after it is poured, and give it a glaze coat of steep roofing asphalt.*

➤ *This makes ice much easier to remove because it does not stick to the asphalt.*

➤ *Take care, however, that the combined weight of this glaze coat and the asphalt used to apply the first layer of the sub-sequent roofing system does not exceed the recommended maximum for proper adhesion.*

➤ *Too much asphalt can cause slippage in hot weather.*

➤ *Remove any moisture from the deck surface.*

➤ *Moisture can cause poor adhesion or skips in the mopping asphalt, which can entrap moisture within the roofing sys-tem. Squeegee all fiberglass ply felts to assure adhesion.*

➤ *Follow the true temperature guidelines for materials and per-sonnel with care. Consider the use of temporary roofs if con-struction schedules require roof applications in cold or rainy weather.*

- In cold weather, the general contractor must be monitored.
- Make sure the project manager coordinates the work of the other trades and installs the nailers, curbs, and other items needed to complete the work in a timely manner.
- If possible, do not let the deck installation get too far ahead.
- While snow is not too hard to remove, ice is difficult.
- When job conditions permit, consider using larger crews to move through the installation rapidly.
- Develop a habit of careful communication with the general con-tractor at all times, not just in cold weather.

- Have regular meetings, discuss any defects in the work of the other subs, and be strict and diligent about change orders.

- Put everything in writing and have it signed. If there is cause for litigation at a later date, a precise paper trail will be of great advantage to your company.

APPLICATION INFORMATION

Flashing

- Flashings are designed to prevent the infiltration of water at any roof area where the membrane is interrupted, terminated, or joins an area or projection having a marked change in slope or direction.

- Membrane interruptions typically occur at gravel stops, curbs, vents, parapets, walls, expansion joints, skylights, drains, and built-in gutters.

- Flashing is generally divided into two categories:
  - Base flashings
  - Counterflashings

- Base flashings are, in a sense, a continuation of the membrane that is turned up on a surface and installed as a separate operation.

- Base flashings are usually of a nonmetallic material such as plastic, or an asphalt-impregnated product.

- Counterflashings or cap flashings can be made of metal, plastic, or impregnated felt. They shield or seal the exposed edges of the base flashing.

- Combining nonmetallic materials and metal in base flashings brings out the best in each material.

- Since the bituminous-base flashings have the same movement capabilities as the roof membrane, they work together as a unit and are recommended as the only type of material for base flashings.

- Because of the rigidity of metal and its extreme movement with temperature changes, its use in base flashing is not recommended.

- Metal cap, or counterflashings installed where there is no possibility of standing water, is acceptable only after composition base flashing is properly sealed.

- Flashing can be applied by hot or cold application methods. Asphalt-coated fiberglass can be applied in a hot mopping of bitumen.

- The advantages of hot-applied flashing are speed, economy, and convenience.

- The proper grade of asphalt must be used.

- Do not use coal-tar pitch or low-softening-point asphalt because flashings embedded in these bitumens sag.

- Secure hot-applied flashings at the top edge with mechanical fasteners spaced 4 to 8 inches on center or by some other positive means to prevent sliding or sagging.

- Use reinforced base flashing in conjunction with a ply of fiberglass felt as a backer felt. Set it in a mopping of hot asphalt to the primed masonry surface.

- The guarantee period to be offered for the roof's life is the guideline that determines the need for a backer felt.

- The cold-application system consists of alternate layers of ply felt set in, and covered with, trowelings of Bestile flashing cement.

- The thinner ply felt conforms readily to the angle between deck and parapet.

- Bestile flashing cement is basically the same as industrial roofing cement, but it contains certain non-asbestos reinforcing fibers.

- This material sets up quite hard compared to remaining plastic.

fastfacts

➤ *Bestile flashing cement provides adhesion to the primed wall and to each ply, enhances interply waterproofing, and provides more time for careful application. Flashing systems of this type set up quite hard, so they do not have the tendency to sag or run under hot temperatures.*

➤ *They are especially practical when no nailing facilities are available. A major benefit to the Bestile flashing system is its compatibility with any bituminous roofing system.*

- Bestile is used almost exclusively for flashing work on vertical surfaces where differential movement is not expected. It should not be used with metal flashing.

- The solid fastening of all flashing accessories to wooden nailers or the substrate is mandatory.

- Unequal movement, due to poorly secured flashing, results in splits in the felt stripping and subsequent leaks.

Flashing Cement

- There are two main types of bituminous flashing cements for cold application.

- Industrial roof cement is a non-asbestos, fiber-reinforced, solvent-cutback, all-purpose plastic cement.

- This cement remains in a semi-plastic state and is used where some movement is expected, such as under a metal roof edge where the differential movement between metal and felt is absorbed in the plastic cement.

- It also is used to repair blisters and to accomplish general roof maintenance by itself and in conjunction with fiberglass-ply felt on horizontal surfaces.

Flashing Details

- There are many flashing designs that the specifier or architect might desire.

- Figures 5.4 through 5.6 illustrate the common designs.

- When bidding on a project, pay close attention to all complex roof joints and flashing details.

- These areas are very common triggers of roof leaks and resultant damage.

- If the details are not drawn in a precise manner, address the issue with the contractor, design anchor, and principal.

- Have acceptable drawings completed, or include a clause in the addendum to the contract that requires a signed inspection slip from the architect during execution.

- Remember that this pause for the architect will add to the job costs.

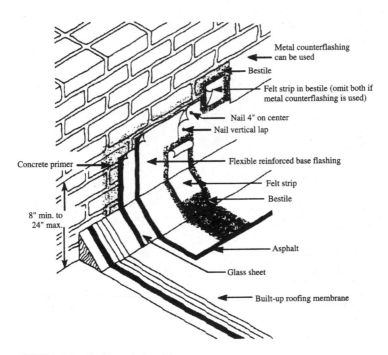

FIGURE 5.4 Flashing design #1.

- It is important to make certain that each of the BUR layers is applied in a diligent, professional manner so the plies are thoroughly welded. This prevents callbacks, promotes word-of-mouth business, and can serve to prevent future lawsuits by disgruntled building owners.

- For example, in Figure 5.4, the roofing felts must extend to the top of the cant.

- Starting just below the point where the base flashing terminates, mop the wall and the roofing felts on the surface of the cant with Type III or IV asphalt that is hot.

- Immediately set the backer felt, if required, into the hot asphalt.

- Smooth the felt to set it firmly into the asphalt.

- Terminate the bottom edge of the backer felt at the bottom edge (base) of the cant.

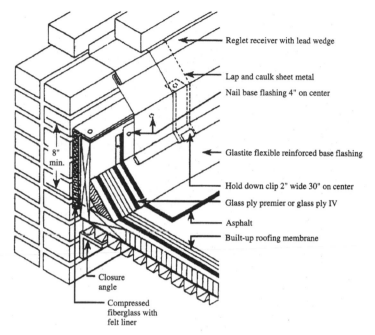

Reglet receiver with lead wedge

Lap and caulk sheet metal

Nail base flashing 4" on center

8" min.

Glastite flexible reinforced base flashing

Hold down clip 2" wide 30" on center

Glass ply premier or glass ply IV

Asphalt

Built-up roofing membrane

Closure angle

Compressed fiberglass with felt liner

FIGURE 5.5 Flashing design #2.

- Starting just above the top edge of the backer felt, mop the wall and the backer felt with hot Type III or IV asphalt.
- Back mop a 6 to 8 foot fiberglass section of flexible reinforced base flashing.
- Holding the upper corners of the base sheet, position its horizontal edge on the roof membrane and roll it into place over the cant strip and up the wall.
- Mechanically fasten the base flashing on 4 to 8 inch centers along the top edge and at the laps.
- Drive fasteners through caps that are at least 1 inch in diameter, unless they have an integral flat cap at least 1 inch across.
- Cover the vertical laps with a 4 inch wide strip of ply felt embedded in and trowled over with ⅛ inch thick layer of Bestile or roofing cement.

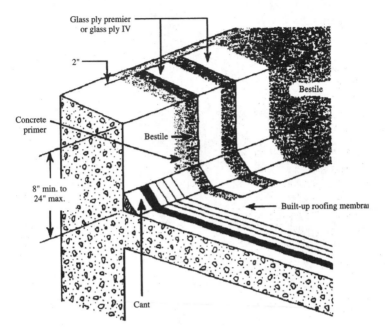

FIGURE 5.6 Flashing design #3.

Applying BUR Penetration Flashing

The penetration expansion joint covers, stacks, supports, and vents on BUR installations are critical to a successful, watertight roof. The following are methods for treating penetrations:

- Expansion joint covers: The type of expansion joint cover that is illustrated, when curb mounted, allows for building movement in all directions (Figure 5.7a).

- Stack flashings provide curbs for accessories that pass through the roof deck.

- A typical example is the vent pipe in Figure 5.7b.

- Also provides curbs for roof fixtures such as skylights or ventilators, and for stacks and conduit.

- Equipment or sign supports above the roof level simplify maintenance, which is very important to the life of the roof (see Figure 5.7c).
- For heavy loads, place continuous supports over the buildings structural supports such as beams or columns.
- Use one-way roof vents (Figure 5.7d) to reduce the moisture vapor content of insulation or certain types of wet-fill poured decks.
- Vents can also be used in conjunction with a ventsulation felt when parapet or roof-edge venting is not practical.
- The placement of vents depends on job conditions, but there should be a minimum of one vent for each ten squares of roof area.

Welding Plies

Built-up roofs are constructed in layers as shown in Figure 5.1.

- Typically, the first layer applied on top of the roof sheathing is insulation.
- Then come plies of roofing felt which are welded when heated. Mopped bitumen melts, and then fuses with the saturant bitumen in the roofing felts.
- Since thorough fusing of the plies is important, correct application temperatures are vital to the creation of a quality roof membrane.
- A high bitumen temperature must be maintained to promote the welding process.
- The drawing in Figure 5.1 illustrates Figure 5.8.
- It is an example of how a built-up roof may be layered.

UNDERSTANDING BITUMENS

- Extended high temperatures can reduce the softening point of asphalt.
- Extended heating can raise the softening point of coal-tar bitumen.
- Bitumens can be kept at high temperatures for short periods without damage.
- Bitumens must be heated to high temperatures to fuse with the felt and give the plies a strong bond.

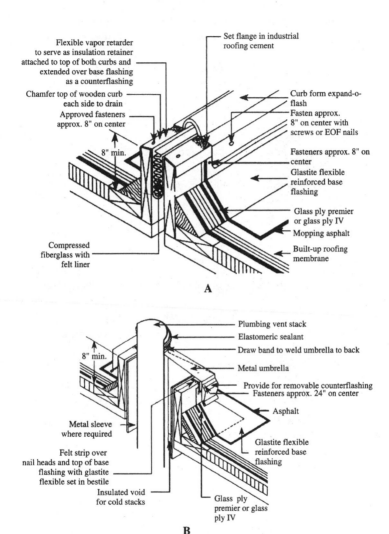

FIGURE 5.7 (a) Expansion joint cover. (b) Stack flashing.

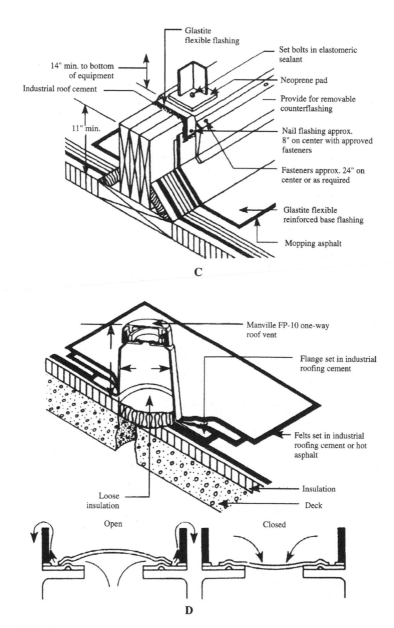

FIGURE 5.7 *(continued)* (c) Equipment or sign support. (d) One-way roof vents.

BUR LAYERS JOBSITE CHECKLIST		
✔	ITEM	NOTES
	Deck - framing & sheathing inspected	
	Insulation - manufacturers instructions followed diligently	
	Bitumen coating - applied with care	
	Insulation - manufacturers instructions followed diligently	
	Bitumen coating - applied with care	
	Roofing felt - Clean, dry, correct weight, applied with care	
	Bitumen coating - applied with care	
	Roofing felt - Clean, dry, correct weight, applied with care	
	Bitumen coating - applied with care	
	Roofing felt - Clean, dry, correct weight, applied with care	
	Bitumen coating - applied with care	
	Roofing felt - Clean, dry, correct weight, applied with care	
	Bitumen flood coating - applied with care	
	Aggregate or mineral surface - applied with care	

FIGURE 5.8 Jobsite checklist for layering of built-up roofing.

- Asphalt is made up of three elements:
  - Saturants are light oils that make the asphalt soft and flexible so that it expands and contracts with the application surface.
  - Asphaltenes provide the body, rigidity, and strength to the application.
  - Resins bond the saturants and asphaltenes and furnish the asphalt with its resilience.
- There are four grades of asphalt:
  - Type I (dead level). This grade of asphalt is relatively susceptible to flow at roof temperatures and has good adhesive and self-healing properties. It is generally used in aggregate surfaced roofs on slopes up to ½ inch per 12 inches.
  - Type II (flat). Type II asphalt is moderately susceptible to flow at roof temperatures. It generally is used on roofs with slopes between ½ inch per 12 inches and 1½ inches per 12 inches.
  - Type III (steep). Steep asphalt is relatively unsusceptible to flow at roof temperatures. It generally is used on roofs with a slope between 1 and 3 inches per 12 inches.
  - Type IV (special steep). This asphalt is resistant to flow at roof temperatures. It is useful in areas where high year round temperatures are experienced. It generally is used on roofs with a slope between 2 and 6 inches per 12 inches.
- The National Roofing Contractors Association (NRCA) adopted Equiviscous Temperature (EVT) definitions to determine flow of bitumens.
- The viscosity (flow) of bitumens must be regarded for tight applications.
- If the EVT information is not furnished by the manufacturer, use the temperature given in Table 5.2 as a guide.
- Coal tar is a by-product of the coking of bituminous coal during steel making (Table 5.3).
- Coal tar offers excellent resistance to aging, chemical breakdown, and moisture.
- Coal tar is known for its resistance to ponded water and its cold flow property—the ability to self-heal cracks and fissures.
- Coal tar's flow limits it to low slope buildings.

TABLE 5.2 Heating Temperatures of Asphalt

HEATING TEMPERATURES OF ASPHALT		
ASTM D-312 TYPE #	ASPHALT TYPE	MAXIMUM HEATING TEMPERATURES (°F)
I	Dead level	475
II	Flat	500
III	Steep	525
IV	Special Steep	525

TABLE 5.3 Types of Coal Tar

TYPES OF COAL TAR		
TYPE & NUMBER	USE	TEMPERATURE RANGES FOR MOPPING & MECHANICAL SPREADERS (°F)
I Coal-tar Pitch	BUR systems with coal-tar saturated felts	360
II Coal-tar Pitch	Damp proofing and membrane waterproofing (below grade & decks)	360
III Coal-tar Bitumen	Bur systems (higher softening point than Type I coal-tar pitch	375

SELECTING FELTS AND BASE SHEETS

- Saturated roofing felts made of either asphalt or coal-tar bitumen can be used as a saturant for organic felts. Organic felts are manufactured from the fiber of paper, wood, rags, or a combination of the three.

- When talking about weight, a felt might be designated as 27 gauge. This means that the unsaturated product weighs approximately 27 pounds per 480 square feet of paper measure and is a nominal 27 mils in thickness.

- The saturated product is generally referred to as a nominal number. For example, No. 15 in a No. 15 asphalt-saturated organic felt describes the nominal, or approximate, weight in pounds of one ply of No. 15 asphalt-saturated organic felt needed to cover one roof square, or 100 square feet.

- Saturated and coated felts are made from organic or inorganic materials. Generally, they are factory-coated on both sides and are surfaced on one or both sides with very fine mineral sand or other release agents to prevent adhesion inside the roll prior to application.

- Impregnated felts are generally lighter in weight than organic felts. They are termed impregnated, rather than coated, because the surface is not completely covered with asphalt. These felts usually provide an open, porous sheet through which vapors can be vented during membrane application.

- A commonly used prepared roofing material is the reinforced flashing membrane, which consists of a glass-fiber base felt that is laminated with cotton or glass-fiber fabric and coated with asphalt.

- Prepared roofing materials are usually packaged in rolls that are one roof square, or 100 square feet, in size.

Prepared Materials

- Prepared roofing materials are saturated and coated felts with talc, mica, sand, or ceramic granules incorporated into their weather surfaces.

- This provides both weather protection and decoration. Prepared roofing materials can be manufactured from organic or glass-fiber base felts or a combination thereof.

Basic Sheets

- No. 30 asphalt-saturated organic felts are medium-duty, unperforated felts that are used as the underlayment for shingle, slate, and tile work and, in some cases, as the base ply in BUR systems.

- No. 40 asphalt-saturated and coated organic-felt base sheets are the base layer for some BUR membranes. These sheets are designed to prevent the intrusion of moisture from below the roof and to minimize the occurrence of ridges and wrinkle cracking. They also are incorporated where the designer requires a tougher or thicker base to help span joints in the deck or substrate immediately below the roof membrane, and where mechanical fastening of a base ply is required.

- Asphalt glass-fiber mat sheet is material composed of a glass-fiber mat that is impregnated and coated on both sides with asphalt. It is available in perforated or non-perforated styles and is designed for use as a base ply in the construction of BUR membranes.

Venting Asphalt-Saturated and Coated Inorganic Sheets

- Several manufacturers make a special, heavy-gauge felt to vent certain roof conditions.

- The venting characteristic of the felt is assisted by mineral granules on the underside of the felt.

- These venting sheets are, in some cases, specified for use over damp or concrete substrates as an underlayment or base ply.

- The sheets can be nailed in place and/or spot-mopped with hot bitumen.

- If employed in a BUR system, these sheets provide for the venting of moisture vapor pressure by lateral or horizontal movement.

Ply Sheets Used Between Coats of Bitumens

- No. 15 asphalt-saturated organic felts are manufactured in two types.

- Perforated felts, which are ply marked for use in two, three, or four layer applications.

- Unperforated felts, which are generally used as the underlayment of shingles and for general building paper usage.

fastfacts

➤ *No. 40 asphalt-saturated and coated organic-felt base sheets are also used for temporary roofing or temporary flashings and as vapor retarders beneath insulation.*

➤ *If a vented base ply does not have a buttonhole attachment and is to be mopped in place, spot mopping is recommended, preferably with a mechanical spot mopper.*

➤ *This prevents solid mopping or strip mopping of the asphalt under the base sheet, which restricts the venting properties of the base sheet.*

➤ *No. 50, smooth-surfaced material is also useful in the construction of temporary roofs and for flashing applications.*

- No. 15 coal-tar saturated organic felts are ply marked for two, three, or four-ply applications and are commonly used in waterproofing specifications and in coal-tar BUR specifications.
- Asphalt-impregnated glass-fiber felts are used in the construction of BUR membranes.
- Asphalt-impregnated glass-fiber felts (Type IV) are slightly heavier than the Type III asphalt-impregnated glass-fiber felts. They have a breaking strength in both longitudinal and transverse directions that is nearly twice that of Type III asphalt-impregnated glass-fiber felts.

Caps and Flashings

- Several caps and flashings are available for use in BUR.
- Wide-selvage asphalt cap.
  - This roll roofing material is composed of an asphalt-saturated felt that is coated on approximately one-half of the weather side.
  - The asphalt-coated surface is then treated with colored mineral granules.
  - It is generally used for two-layer applications over a base ply or as succeeding sheets over the initial plies of a BUR system.

- No. 90 mineral-surfaced roll material
  - This product is manufactured either with no selvage or with a 2 inch selvage, depending on the manufacturer.
  - It is supplied with various ceramic-colored granule surfacings, and is used as an organic roll roofing material and specified for a variety of flashing applications.
- No. 50 smooth-surfaced material
  - This roll asphalt product is used in some roof membrane specifications as a heavy-duty base sheet.
- No. 60 mineral-surfaced material
  - This roll-roofing product is an organic felt that is saturated with a mineral-stabilized asphalt compound and surfaced with talc or mica.
  - It is used when inexpensive roll roofing is desired, as well as for flashings and temporary roofs.
- Inorganic cap sheet
  - This roll material is composed of a heavyweight, inorganic felt that is saturated and coated on both sides with asphalt and then surfaced on the weather side with various colored mineral granules.
  - It is used as a top or cap surface over other BUR plies, for flashings, for granular-surfaced lightweight roof system constructions, and for decorative purposes.
- Reinforced bituminous flashing.
  - This material is a heavy saturated felt that is reinforced with glass fiber.
  - The material is surfaced with fine mineral matter.

APPLYING BUR MATERIALS

The proper application of roofing materials is as important as the materials themselves. The following are recommended guidelines for the application of all roofing materials.

Application Guidelines

- Never use wet or damaged materials.

- Never apply any roofing materials during rain or snow or to wet surfaces.

- Moisture trapped within the roofing system can cause severe damage to the roofing membrane and insulation.

- Never mop more than 4 feet ahead of the roll in temperatures below 50°F.

- Observe EVT (Equiviscous Temperature) guidelines for asphalt application.

- When using mechanical felt-laying equipment, be sure all orifices are open.

- Lightly squeegee all fibrous felts into the hot asphalt.

- Roll all felts into the hot asphalt.

- Take special care when applying any roofing felts in cold weather.

- Check the temperature of the asphalt at the mop or spreader to determine if it is at the proper temperature.

- Do not mix different grades of asphalt or dilute asphalt with any materials.

- Do not use coal-tar pitch or bitumen with any MBR products.

- Do not use cutback asphalt cements under any MBR products.

- Heat the asphalt according to the manufacturer's recommendations.

fastfacts

➤ *The use of the mastics over the top of the product to strip in or cover nail heads is acceptable, however, MBR cement products are preferred.*

➤ *Always check the temperature of the asphalt at the kettle AND at the point of application.*

➤ *Phased construction can result in the felts slipping because of excessive amounts of asphalt between the phased felt plies.*

➤ *Blisters, due to entrapment of moisture, are also a common problem, as is poor adhesion due to dust or foreign matter collecting on the exposed felts of an incomplete roofing system.*

- Use the proper grade asphalt as specified by the membrane manufacturer.

- A good guideline to follow is to use the softest grade of asphalt commensurate with the slope and climate conditions.

- Install water cutoffs at the end of each day's work to prevent moisture from getting into and under the completed roof membrane.

- Be sure to remove the cutoffs prior to continuing the application process.

- Install the complete roofing system all at one time.

Figure 5.9 is a handy checklist for your BUR jobs. It covers all of the above information.

INSTALLING BUR ROOFS

- Manufacturers offer a large variety of information about their roofing products, and they provide a good deal of technical overview to the roofing community.

- When you have any doubt about an installation, go to the manufacturers of the products, the architect for the job, the NRCA, and check out the Internet.

- A little time spent getting the work product correct during the roofing project will reduce callbacks and the potential for future lawsuits.

Temporary Roofs

- There are times when a project needs to be closed-in when the weather is not conducive to good roof work, or the roof area might have to be used as a work platform during construction.

- In these situations, it is recommended that you keep in mind the following requirements:

  - The components of temporary roof systems are all asphalt-related materials.

  - The roof deck must be secured to the structure in such a way that it is able to support the designed live- and dead-load factors required for the building.

✓	Item
	Never use wet or damaged materials.
	Never apply any roofing materials during rain or snow, or to wet surfaces. Moisture trapped within the roofing system can cause severe damage to the roofing membrane and insulation.
	Review the application guidelines in this book.
	Never mop more than 4 feet ahead of the roll in temperatures below 50°F.
	Observe EVT guidelines for asphalt application.
	When using mechanical felt-laying equipment, be sure all orifices are open.
	Lightly squeegee all fibrous felts into the hot asphalt.
	Roll all felts into the hot asphalt. Do not "fly" felts into the asphalt.
	Take special care when applying any roofing felts in cold weather. Check the temperature of the asphalt at the mop or spreader to determine if it is at the proper temperature.
	Do not mix different grades of asphalt or dilute asphalt with any materials.
	Do not use coal-tar pitch or bitumen with any MBR products.
	Do not use cutback asphalt cements under any MBR products. The use of the mastics over the top of the product to strip in or cover nailheads is acceptable; however, MBR cement products are preferred.
	Heat the asphalt according to the manufacturer's recommendations. Check the temperature of the asphalt at the kettle and at the point of application.
	Adhere to the guidelines for heating asphalt in this book.

FIGURE 5.9 Applying BUR materials.

✓	Item
	Use the proper grade asphalt as specified by the membrane manufacturer. A good guideline to follow is to use the softest grade of asphalt commensurate with the slope and climate conditions.
	Install water cutoffs at the end of each day's work to prevent moisture from getting into and under the completed roof membrane. Be sure to remove the cutoffs prior to continuing the application process.
	Heed the specific cold-weather application procedures given in this chapter.
	Install the complete roofing system at one time. Phased construction can result in the felts slipping because of excessive amounts of asphalt between the phased felt plies. Blisters, due to entrapment of moisture, are also a common problem, as is poor adhesion due to dust or foreign materials collecting on the exposed felts of an incomplete roofing system.
	During the asphalt setting time, which can be as long as 45 minutes, it is essential that traffic be minimized so that the interply mopping is not displaced.
	Comply with published safety procedures for all products. Refer to appropriate instructions and product safety information.

FIGURE 5.9 *(continued)*   Applying BUR materials.

- A temporary roof is a roof membrane constructed of a minimum amount of material and placed over a building to permit limited construction to progress during severe weather conditions.

- It is not expected to exhibit the same watertight integrity as a new roof.

- Put other trades people on notice that care must be exercised when working on or over a temporary roof membrane, or when placing penetrations through a temporary roof membrane.

- Raise heating, ventilation, and air conditioning (HVAC) units above the plane of the temporary roof to allow for flashing beneath the HVAC units.

- When putting a temporary roof over a nailable deck, install two plies of No. 15 asphalt-saturated, perforated organic felt.

- Start at the low point of the roof deck, such as a valley or drip edge.

- Overlap these plies 19 inches, leaving an exposure of 17 inches, and nail.

- Embed the plies into a fluid, continuous application of asphalt.

- Apply the asphalt so that felt does not touch felt and no asphalt is applied to the substrate.

- Nail the back half of each sheet and cover it with a fluid, continuous application of asphalt.

- Embed the next ply of felt into the covered sheet.

- All plies of felt are to be broomed into place as they are applied to aid adhesion.

- Heat the asphalt in accordance with EVT standards and apply within the temperature range (EVT ± 250°F).

- When applying a temporary roof over a steel deck:

  - Use one thin layer of perlitic-board roof insulation.

  - The board should have a minimum thickness sufficient to span the flutes of the steel deck.

  - Mechanically fasten the board to the deck with approved mechanical fasteners.

  - Install two plies of No. 15 asphalt-saturated, perforated organic felt over the primed concrete deck.

- Follow these same guidelines for installing a temporary nailable deck.

- Prime the concrete with a coating of asphalt primer applied at the rate of 74 gallon of primer per 100 square feet.

- Figure 5.10 is a checklist for temporary built-up roofs.

✓	Item
	The components of this temporary roof system are all asphalt-related materials.
	Secure the roof deck to the structure in such a way that it is able to support the designed live- and dead-load factors required for the building.
	A temporary roof is a roof membrane constructed of a minimum amount of material and placed over a building to permit limited construction to progress during inclement weather conditions. As such, it is not expected to exhibit the same watertight integrity as a new roof.
	Put other tradespeople on notice that care must be exercised when working on or over a temporary roof membrane or when placing penetrations through a temporary roof membrane.
	Raise heating, ventilation, and air conditioning (HVAC) units above the plane of the temporary roof to allow for flashing beneath the HVAC units.

FIGURE 5.10 Temporary built-up roofs checklist.

One-Layer Insulation Systems

- Apply a base ply to the nailable roof deck following the instructions for the two-layer system.

- Lay the first layer of insulation in hot asphalt or adhesive over the base ply.

- The long joints of the insulation boards should be laid in a continuous straight line with end joints staggered.

- If hot asphalt is used, heat the asphalt in accordance with EVT standards and apply within the temperature range.

- Figure 5.11 shows the attachment of insulation to a steel deck.

- Figure 5.12 illustrates two and one-layer installation over a non-nailable concrete roof deck.

- Figure 5.13 shows loose-laid insulation, which can be installed on any surface.

- Generally, the long dimension of the insulation boards runs perpendicular to the roof slope.

- Use only enough mechanical fasteners or adhesive to hold the insulation boards in place during the installation of the roofing membrane.

- Attach the second layer of insulation, if used, to the first layer with compatible adhesive or hot asphalt.

- Roof insulation can be applied on inclines up to 6 inches per 12 inches, but it must be acknowledged that nailer usage violates the vapor retarder as a result of fastener penetrations, as does the perimeter nailing of insulation for uplift resistance.

- The National Roofing Contractors Association (NRCA) recommends that consideration be given to mechanically fastening or spot adhering the first layer of insulation to the roof deck under loose-laid systems.

- In an alternative procedure for nailable roof decks, both layers can be mechanically fastened to the deck at the same time.

- Roof insulation can be used on pitches up to 6 inches per 12 inches of rise. However, you must remember that using nailers violates the vapor retarder as a result of the fastener penetrations, as does the perimeter nailing of insulation for uplift resistance. Be sure to consult with the architect and building owner for any unique situations and get changes in writing.

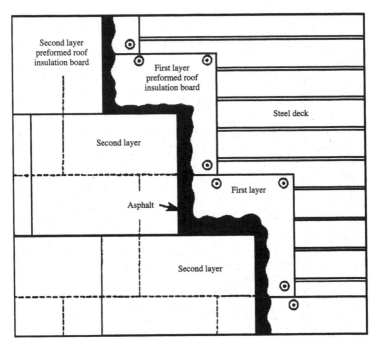

FIGURE 5.11 Attaching insulation to steel decks.

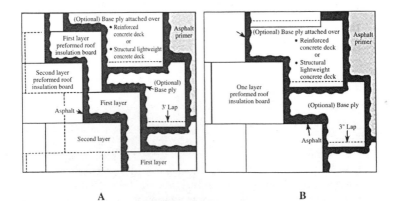

FIGURE 5.12 (a) Two-layer. (b) One-layer BUR over concrete deck (without nails).

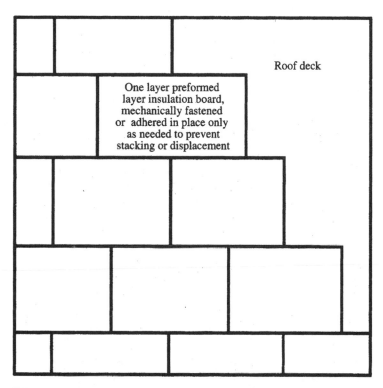

FIGURE 5.13   Loose-laid insulation on any roof surface.

- When the deck incline is such that roofing felts must be nailed (2 inches and over for smooth-surfaced roofs and 1 inch and over for gravel and mineral-cap sheet-surfaced roofs), use wooden nailing strips (4 inch minimum nominal, 3½ inch minimum actual width) at the ridge and at the approximate intermediate points.
- Attach nailing strips to the deck that are the same thickness as the insulation and at least 3½ inches wide.
- Lay out at right angles to the incline to receive the insulation and retain the nails securing the felts.
- Run the felts parallel to the incline, at right angles to the nailers.
- Nails must have at least a 1 inch diameter cap. When capped nails are not used, place fasteners through caps that have a minimum diameter of 1 inch.

- For three, four, and five-ply roofs, locate a nail at each nailer and sheet application space ¾ inch from the leading edge of the felt.

- For cap sheet application, locate one nail at each nailer and space approximately ¾ inch from the leading edge of each ply felt.

- Figure 5.14 shows the pattern for the nailers and spacing for four-ply systems. The termination of a continuous cap sheet must occur at a nailer.

- At points of termination, locate five nails at each nailer (Figure 5.15).

- Space the first nail ¾ inch from the leading edge of the cap sheet and the remaining four nails approximately 8½ inches on center (Table 5.4).

- Stagger the nails across the width of the nailer to reduce the chance of the cap sheet tearing along the nail line.

- It is recommended that each succeeding row of fasteners be staggered from the preceding row. For example, nail one row of fasteners on the lower half of the nailer board, then nail the succeeding felt on the upper half to avoid tearing the felts.

- Table 5.5 shows nailer spacing for asphalt/cap sheet, asphalt/ smooth-surfaced, and asphalt/gravel surfaced roofs.

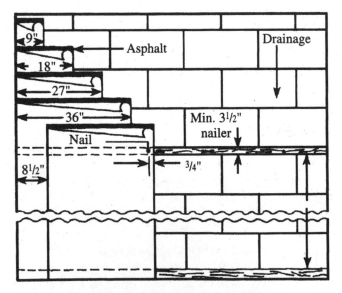

FIGURE 5.14 Pattern for nailer and spacing with four-ply system.

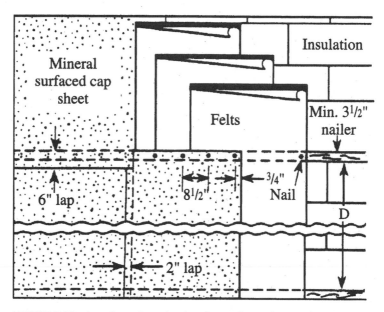

FIGURE 5.15 At points of termination, locate five nails at each nailer.

TABLE 5.4 Nailer Spacing

Nailer Spacing			
Incline (in.)	Smooth	Gravel	Cap sheet
0–1	Not required	Not required	Not required
1–2	Not required	20' face to face	20' face to face
2–3	20' face to face	10' face to face	10' face to face
3–4	10' face to face	Not recommended	4' face to face
4–6	4' face to face	Not recommended	4' face to face

TABLE 5.5 Nailer Spacing for Various Roof Types

Nailer Spacing for Various Roof Types

Incline (in.)	Nailer spacing (D)	Asphalt type
Asphalt/cap sheet roofs		
0–1/2	Not required	II
1/2–1	Not required	III
1–2	20' face to face	III
2–3	10' face to face	III
3–6	4' face to face	IV
Asphalt/smooth-surfaced roofs		
0–1/2	Not required	II
1/2–1	Not required	II
1–2	Not required	III
2–3	20' face to face	III
3–4	10' face to face	IV
4–6	4' face to face	IV
Asphalt/gravel-surfaced roof		
0–1/2	Not required	II
1/2–1	Not required	III
1–2	20' face to face	III
2–3	10' face to face	III

INSTALLING BUR MEMBRANES

- Various numbers of plies can be used in the construction of a BUR system.
- Figure 5.16 illustrates the use of three, four, or five-ply membranes over a structural deck (without insulation) that can receive and adequately retain nails or other types of mechanical fasteners as recommended by the deck manufacturer.
- Examples of such decks are wooden plank and plywood.

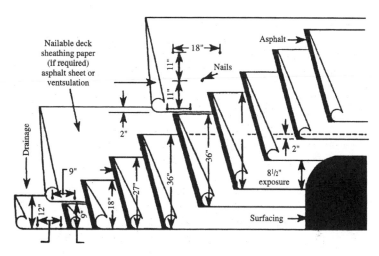

FIGURE 5.16 Three, four, and five-ply membranes.

fastfacts

➤ *The procedure in Figure 5.11 is not for use directly over gypsum, lightweight insulating concrete decks either poured or precast, or over fill made of lightweight insulating concrete.*

➤ *Be sure to prime poured and precast concrete decks.*

Figure 5.17 illustrates a four-ply, gravel-surfaced, fiberglass mat BUR application that can be used over any type of structural deck, which is not nailed and that offers a suitable surface to receive the roof.

SURFACING BUILT-UP ROOFS

- BUR systems consist of multiple layers of bitumen, which are the waterproofing medium, and various forms of roofing felts, which are the reinforcements.
- Each type of base, cap, and ply felt is designed to perform a specific function in the roofing system.
- Variations of materials permit the installation of BUR systems over many different substrates.
- Bitumen is available in various grades to accommodate conditions that might be encountered, e.g., slopes, products, and weather conditions.

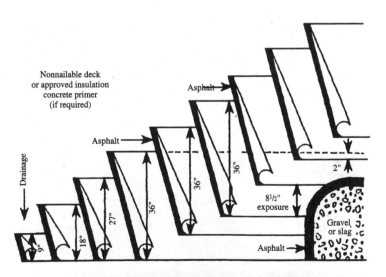

FIGURE 5.17   Four-ply BUR with fiberglass mat, gravel surfaced.

THE THREE BASIC TYPES OF BUR SYSTEMS

Smooth Systems

- Lightweight.
- Easy to inspect and repair.
- Absence of gravel facilitates fast visual inspection.
- Patching is simple and fast.
- Maintenance is much easier than other BURS.
- No need to remove the old smooth-surfaced membrane when re-roofing.
- When removal is needed, it is simpler than with other BURS.
- Smooth-surfaced roofs are cost effective.

Gravel Systems

- Require a flood coat of approximately 60 pounds when applying the gravel.
- Typically, more durable than smooth surfaced BUR systems.
- The gravel helps the flood coat resist the aging effects of the elements.
- The gravel also helps stabilize the flood coat for heavier pours of bitumens.
- Offer improved fire resistance from the aggregate.
- Added protection from penetrating forces like hail.
- Generally limited to slopes of 3 inches or less to minimize gravel loss and membrane slippage.

fastfacts

➤ *Lightweight systems are typically less than one-third the approximate 400 to 700 pounds per square of gravel-surfaced roofs.*

Mineral Systems

- Mineral-surfaced cap sheets form the last or visible ply.

- The inorganic mat is coated with weather-grade asphalt into which opaque, noncombustible, ceramic-coated granules are embedded.

- The resulting sheet yields a roof that enhances the appearance of the building it protects.

- The ceramic form, factory-applied surfacing helps the underlying bitumen resist weathering and aging.

- Unlike coatings, these granules are not-surfaced, which means simple maintenance. Mineral system cap sheets offer improved fire resistance and reflective properties.

6

SINGLE-PLY ROOFS

Factory produced membrane systems first developed in the European countries' found their way into the United States market in the 1960s. They now account for around 40 percent of the flat roof market. They continue to grow in popularity because they have proven to be sturdy even under the rough tests a roof receives.

- A prime feature of single-ply systems is that they do not require the products, equipment, and intense labor required for built-up, hot-mopped roofs.
- Single-ply systems are based on rolled sheets of synthetic compounds.
- They are manufactured in factories where attention to quality control can be thorough.
- Single-ply products are tough, relatively easy to install, and can last a long time.

SPRI (SINGLE-PLY ROOFING INSTITUTE) CLASSIFICATIONS OF THERMOSETS

- Thermoplastics
- Modified bitumens

FIGURE 6.1 A single-ply installation.

Casting Thermosets

- Single-ply thermosets are polymer products that cannot be changed once the sheet material is cast.
- Two types of thermosets are used for roofing:
  - Vulcanized (hardened rubber or rubber-like material), or cured.
  - Elastomer (natural or synthetic polymer with elastic properties like rubber) and non-vulcanized, or non-cured, elastomer.
- Advantages of elastomeric roofing:
  - Rugged and long-lasting.
  - Cost savings.
  - Environmentally friendly.
- Adaptable to many conditions, conservation, the substitution of materials, and adaptability to a wide range of roof configurations

FAVORABLE QUALITIES OF SINGLE-PLY ROOFS

- Extension or elongation of certain systems at room temperature can be as high as 700 to 800 percent.

fastfacts

➤ *Single-ply roofing systems can even elongate and accommodate movement in the substrate. Elastomeric membranes can bridge joints and cracks in the substrate without cracking and splitting, provided they are not bonded, or they are reinforced at these locations.*

➤ *The ability of the membrane to remain flexible at low temperatures is another feature.*

➤ *Discuss these uses carefully with the manufacturer's tech staff, the architect, and the owner before proceeding.*

- Some elastomeric membranes remain flexible at temperatures as low as −50°F, whereas conventional bituminous membranes become brittle within a range of approximately 0°F to 45°F.

- Some elastomeric membranes retain their ability to elongate at low temperatures, although the elongation is reduced from that at room temperature, or approximately 68°F.

- Some elastomeric roofing systems weigh less than 10 pounds per 100 square feet of roof area. This is a minimal weight compared to that of smooth-surfaced bituminous systems, which weigh approximately 150 pounds per square foot.

  - Single-ply comes in a wide variety of configurations.
  - Domes
  - Barrels

- Hyperbolic paraboloids.

- Let the architects and building owners know that you can handle most shapes.

- A variety of colors.

- Colors can also be reflective.

- Colors can provide lower roof temperatures.

Vulcanized Elastomers

- The two most popular vulcanized elastomers:
  - Ethylene propylene diene terpolymer (EPDM).
  - Chlorinated polyethylene (CPE).

Ethylene Propylene Diene Terpolymer (EPDM)

- Compounded from rubber polymer "rubber roofing."
- The American Society for Testing and Materials (ASTM) classifies this material as an M class polymer.
- EPDM single-ply roofing systems features:
  - Long-term weatherability, including excellent resistance to temperature extremes, sunlight, ozone, and moisture.
  - Ease, speed, and cleanliness of installation.
  - Flexural stability enables EPDM membranes to accommodate roof deck movement and displacement.
  - Compatibility with a wide variety of polystyrene insulation products.
  - Ease of maintenance.
  - Proven long-term performance.
  - Various application techniques, such as ballasted, fully adhered, and mechanically fastened, allow EPDM roofing systems to be applied to virtually any roof surface: flat, spherical, curved, or slanted.
  - When the roof deck is moisture-free, EPDM can be installed over existing roof membranes in reroof situations.
  - Code and standardization progress by the Rubber Manufacturers Association (RMA), Underwriters' Laboratories, Inc. (UL), and Factory Mutual (FM) contribute to the demand.

Securing Systems

- Ballasted systems:
  - Most common for rubber-based, single-ply roofs.
  - The thermal insulation and rubber membrane are laid over the roof deck loosely.

- They are then covered with ballast, which is usually round, washed river rock as specified by ASTM.
- The major advantages of ballasted installation are:
  - Cost
  - Ease of installation.
  - UL, Class A fire rating.
  - Separation of the membrane from the deck allows for maximum independent movement.
- Fully adhered systems:
  - They are the second most popular application method (Figure 6.2).
  - Ideal for contoured roofs.
  - Excellent for sloped surfaces that cannot withstand the weight of a ballasted system.
  - Ideal for reroofing applications over existing material.
  - Fully adhered membranes are completely bonded to the substrate using contact adhesives.
- The major advantages include:
  - Light weight
  - Durability
  - Ease of maintenance
  - Aesthetically clean, smooth appearance
- Mechanically-fastened systems:
  - The membrane is laid over the substrate loosely.
  - The membrane is then anchored to the deck using fasteners.
  - Many types of mechanically fastened systems are available.
  - Most common are those attached directly to the deck.
  - Nonpenetrating systems are also available.
  - Mechanically fastened systems are lightweight, easy to install, and relatively inexpensive. As discussed later in this chapter, a recent development in EPDM is a sheet that can be heat-welded.

Chlorinated Polyethylene (CPE)

- CPE exhibits properties of both vulcanized and non-vulcanized elastomers.

FIGURE 6.2 Apply adhesive on a fully adhered EPDM roof.

- CPE is manufactured as a thermoplastic, but over time it cures as a thermoset.
- Most CPE systems are hot-air or solvent-welded.
- Features and benefits of CPE are discussed in greater detail in the non-vulcanized, uncured elastomers section.

Polychloroprene or Neoprene

- Neoprene is a generic name for polymers of chloroprene.
- The first commercially produced synthetic rubber.
- Resistant to:
  - Petroleum oils
  - Solvents
  - Heat
  - Weathering
- Comes in sheet and liquid-applied forms.

- Available in:
  - Weathering grade—black.
  - Non-weathering grades—light-colored, must be protected from sunlight, normally by applying a coating of chlorinated polyethylene.
- Neoprene formulations are no longer widely used for roofing.

Non-Vulcanized, Uncured Elastomers

- Non-vulcanized elastomers can be heat-welded during the initial installation.
- Non-vulcanized elastomers include:
  - Chlorosulfoned polyethylene (CSPE)
  - Chlorinated polyethylene (CPE)
  - Polyisobutylene (PIB)
  - Nitrile alloy with butadiene-acrylonitrile copolymers (NBP)

Chlorosulfoned Polyethylene (CSPE)

CSPE, sold under the DuPont trademark Hypalon, is a polymer that has enjoyed increased popularity over the years because of its attractive white appearance and energy-efficient, heat-reflective properties. CSPE is an elastomeric material with typical elongations range from 200 to 400 percent, depending on the type and amount of reinforcing fillers present in the compound. This means that CSPE can stretch far beyond its original length and return to its original configuration without loss of structural integrity. A unique feature of CSPE is that it is manufactured as a thermoplastic, but over time, it cures as a thermoset. This means that most CSPE systems are typically hot-air or solvent-welded.

The advantages of welded seams include relatively quick procedures, good strength, and the lack of additional seaming material. Only a handful of CSPE systems are installed using the ballasted technique popular with other rubber-based roofing systems because most building owners prefer not to cover the attractive white membrane. In fact, CSPE systems are often specified because of their attractive white exterior. The two main methods of application for CSPE rubber-based, single-ply systems are mechanically fastened and fully adhered.

- Mechanically fastened:
  - The membrane is loosely laid over the substrate.
  - Then, it is anchored to the deck using fasteners. Many types of mechanically fastened systems are available. The most common are those that attach directly to the deck. Nonpenetrating systems are also available. Advantages include the fact that the systems are lightweight, easy to install, and relatively inexpensive.
- Fully adhered:
  - Ideal for contoured roofs or sloped surfaces that cannot withstand the weight of a ballasted system.
  - Fully adhered membranes are completely bonded to the substrate with contact adhesives.
  - The major advantages include the fact that these systems are lightweight, durable, and easy to maintain.
  - They also have an aesthetically clean and smooth appearance.

Chlorinated Polyethylene (CPE)

- CPE is manufactured with pigment and processing aids, which serve as release agents and antioxidants.
- The majority of today's CPE roof membranes are offered in an uncured composition and are reinforced with a polyester scrim by individual roofing manufacturers.
- Standard thicknesses are 40 to 48 mils.
- Both CPE and CSPE normally are formulated without plasticizers because of their inherent flexibility as an elastomer.
- As closely related elastomers, CSPE and CPE share similar behavioral characteristics.
- An aged CSPE, however, cannot be heat-welded. Adhesives must be applied if field repairs are necessary.
- The methods of application are the same for CPE and CSPE.

Polyisobutylene (PIB)

- PIB is usually a 60-mil membrane made from Polyisobutylene, a synthetic rubber polymer, and pigments, and fillers.
- The underside of the membrane is generally laminated with a 40-mil, needle-punched, nonwoven, rot-proof polyester fabric.

fastfacts

➤ *The fact that the system is unballasted can eliminate the need to structurally reinforce the building before it is reroofed.*

➤ *PIB also minimizes the deck load and eliminates the logistics of adding 1000 to 1200 pounds of gravel for every 10 foot square area.*

- The membrane is finished with a 2 inch wide, self-sealing edge material that is protected by a strip of release paper.
- PIB is compatible with hot asphalt.
- It shows excellent resistance to weathering, radiant heat, and ultraviolet (UV) light.
- It is used primarily as a final waterproofing membrane over existing flat or low-sloped roof assemblies.
- PIB can also be used as a waterproofing membrane for new-construction roof assemblies.
- PIB is a lightweight system that requires no ballast.
- It can be installed quickly and economically.
- PIB passes UL's Class A fire rating and FM's I-90 wind-resistance rating.
- Advantages of PIB:
  - The self-sealing edge offers assured seam strength.
  - Long-term waterproofing.
  - Competitive installation costs—faster application per worker hour and no need for special expensive equipment.

Nitrile Alloys (NBP)

- Thermoset elastometers compounded from butadiene-acrylonitrile copolymers, nonvolatile polymeric plasticizers, and other patented ingredients.
- Made by coating the compound on a heavy-duty polyester fabric.

- They range in thickness from 30 to 40 mils.
- Seams are hot-air welded.
- NPB is recognized for its weather protection and waterproofing capabilities.
- Exhibits good chemical resistance and low-temperature flexibility.

HEATING THERMOPLASTICS

- When heat is applied to a thermoplastic, it becomes more pliable and heat-weldable.
- When returned to ambient temperature, it regains its original properties.
- This process can be repeated again and again with the same results, which explains why seaming is so excellent and easily done with a thermoplastic.

   The two most common chlorinated hydrocarbon thermoplastics are polyvinyl chloride (PVC) and Chlorinated Polyethylene (CPE).

Polyvinyl Chloride (PVC)

- PVCs are among the most versatile thermoplastics for industrial and commercial applications.
- PVC is synthesized from vinyl chloride and is a member of a larger group of polymers designated as vinyls. These polymers are composed of intertwined molecular chains. This, in part, is what gives PVC its unusually good physical properties.

fastfacts

➤ *Thermoplastic roofing systems tend to be light in color, which can add value in terms of aesthetics.*

➤ *They are especially popular in multi-tiered roofing that can be seen from above by building occupants or neighbors.*

- PVC is one of the easiest materials to use.
- In its uncompounded state, PVC is a rigid material.
- When blended with plasticizers, PVC becomes soft and pliable.
- The use of the proper plasticizers enables membranes based on PVC to be used over a wide temperature range without substantial change.
- Since PVC is a plastic and not a rubber, it is unaffected by ozone.
- PVC is an excellent choice when high performance and economy of cost are the primary factors in roof selection.
- PVC is forgiving. Installation mistakes can be easily corrected and alterations, such as the addition of air-conditioning systems at a later date, are easily accomplished.
- PVC conforms to nonstandard details with exceptional ease.
- Seams have real integrity.
- Thermoplastic roofing membranes can be welded together with heat or solvents.
- Once welded, they develop bond strengths that equal or surpass the strength of the base material.
- There are three different methods of PVC installation.
- The simplest installation procedure is to loosely lay the material on the substrate.
- Attach the membrane only around the perimeter of the roof and at any penetrations.
- The insulation does not need to be fastened in place, as the roofing system is ballasted to resist wind uplift.
- Since the membrane is not attached to the substrate, stresses in the substrate are not transferred to the membrane.
- Gravel ballast must be of sufficient size and free of sharp edges, so that it does not puncture the membrane.
- The partial bonding method uses elements of both the loose-laid and fully adhered systems.
- Insulation material beneath the membrane must be firmly fastened to the substrate.
- Then mechanically fasten round plates firmly to the substrate in a predetermined pattern and spacing. Finally, bond the membrane to the pattern.
- Partial bonding allows the membrane to float free over a crack or joint in the substrate.

- This distributes stress in the membrane between adhered areas.
- It is not intended to serve as, or replace an expansion joint.
- On partially adhered systems, there are voids between membranes and substrates.
- If moisture enters the void and vapor pressure builds, blistering can occur.
- The fully adhered method completely fastens the roofing system to the roof deck.
- For sheet systems, total bonding is usually achieved by applying an adhesive.
- The substrate for adhesive-bonded systems must be smooth, clean, dry, and free from dust, dirt, grease, oil, wax, and loose particles.
- Many manufacturers recommend that the substrate be primed before adhesive is applied to achieve greater bond.
- Contact adhesives are typically used in total bonding.
- They are applied to the top surface of the substrate and the bottom of the sheet.
- Some sheets are self-adhering, however, so that the bottom of the sheet adheres to the substrate without an adhesive.
- The fully adhered method is suited to reroofing projects when the old roofing is not removed.
- It can also be used on steep roofs that cannot contain ballast and where partial bonding allows the membrane to sag between fastening points.
- Mechanically fasten a layer of hardboard or insulation to the deck, since PVC is incompatible with asphalt or coal tar.
- The system designer must consider all roofing materials when specifying PVC.
- Common roof materials, such as asphalt and coal-tar pitch, cannot come in contact with PVC single-ply roofing. If they do, the PVC membrane may fail.
- All PVC roofing manufacturers stress that the membrane must be completely isolated from any contact with bituminous materials.
- Even fumes from coal tar must be avoided.
- Bituminous materials have the effect of leaching the plasticizers out of the PVC, leaving it weak and brittle.
- Polystyrene insulation must not come in direct contact with PVC, since it can rapidly extract the plasticizers from the membrane.

- The treatment used for wooden blocking and nailers must be carefully considered because only waterborne wood preservatives can be used.

Chlorinated Polyethylene (CPE)

- Thermoplastic elastomers, such as CPE, are unique in that they have elastomeric properties, rubber-like elasticity, and are easy to install, like PVC.
- CPE can be applied in the same manner as PVC.
- Welds can be made with heat or solvent, and detailing is accomplished easily because of the thermoplastic-like properties of CPE.
- It has most of the desirable performance properties of a true rubber and easily conforms to any subsequent alteration to the roof.
- It seams almost as easily after being in place several years as when initially installed.
- Elastomers exhibit the qualities of both rubber and thermoplastic.
- CPE is resistant to acetic acid, asphalt, bleach, chlorine, coal tar, fuel oils, animal fats and oils, fertilizer, sulfuric acid (acid rain), and a broad range of other corrosives.
- CPE has superior resistance to weathering, UV radiation, ozone, and microbiological attack. In addition, CPE has good fire resistance qualities.

MODIFIED BITUMEN ROOFING

Modified bitumen roofing (MBR) systems have their technical roots in Italy, France, and Germany. They first appeared on the American market in 1975.

- They consist of asphalt, a conventional base material, and any one of a number of polymeric modifiers. MBR membranes bridge the gap between conventional hot-applied BUR and nonconventional single-ply membranes.
- MBR systems cover the full range of in-place system specifications.
- These systems currently are being applied using loose-laid, partially attached, and totally adhered methods, and as protected membrane systems.

fastfacts

➤ *CPE offers the best chemical resistance of any conventionally available, single-ply roofing material.*

➤ *The requirements for deck selection, roof insulation, vapor retarder application, drainage, and flashing details that apply to conventional hot-BUR applications are generally applicable to most MBR systems as well.*

➤ *This makes the selection of an MBR system particularly attractive to the roofing contractor who might be skeptical of other single-ply systems' application requirements and techniques.*

➤ *When applying MBR over existing roofs, make sure the substrate is not badly deteriorated or heavily moisture-laden.*

- MBRs can be applied over a wide range of roof decks and roof insulations.
- Installation methods for MBR systems are also versatile. They can be applied by hot-asphalt application, heat-welding or torching, or cold adhesive.
- Some systems have self-adhering proper ties.
- Both the SBR modified and APP modified bitumen systems produce good lap-joint strength.
- Lap-joint construction for the APP modified membranes is usually accomplished by torching.
- SBR and SBS modified membranes are normally adhered using hot asphalt as the adhesive.
- They can also be torched, although the APP modified systems are more conducive to torching.
- The APP modification allows the bitumen material to flow easily when heat is applied directly, thus the term *torching felt*.

- As with BUR membranes, the MBR membrane consists of layers of modified asphalt that waterproof a reinforcing material.

- The reinforcing in MBR might be glass-fiber mats, polyester scrim, or a combination of the two. Each type of reinforcing material imparts different properties to the membrane. The manufacturing process usually dictates the location of the reinforcing material within the membrane.

- Glass-fiber reinforcement material is generally placed close to the top of the mat to serve as a wearing surface. It resists foot traffic and UV degradation and provides a fire rating.

- MBR membranes range in thickness from 40 to 60 mils.

- Most surfacing consists of granules provided by the manufacturer. This granule surfacing does not add much weight to a system.

- Other coverings include manufacturer-installed metals or applied coatings of acrylics, asphalt emulsions, or fibrated aluminum.

- Fibrated aluminum is the most popular because of its reflective properties. The differential movement of metal applied over asphalt is solved by fabricating tiny expansion joints into the metal.

- UV resistant APP can be left uncoated.

- It usually is coated to promote longer life.

- Coatings include:
  - Fibrated aluminum
  - Acrylics
  - Asphalt emulsions

- Mineral surfacings are available, too, but generally are used only for their aesthetic effect.

- Surfacings are also used to achieve fire ratings.

- MBR systems are generally applied in either a single-ply or multi-ply application (Figure 6.3).

- Organic or fiberglass felts are typically used as the base ply.

- Heat welding and hot-asphalt mopping are typical.

- Use MBR systems for new construction application as well as reroofing jobs.

- Due to their superior load-elongation behavior, MBRs are frequently applied directly over existing conventional roofing materials.

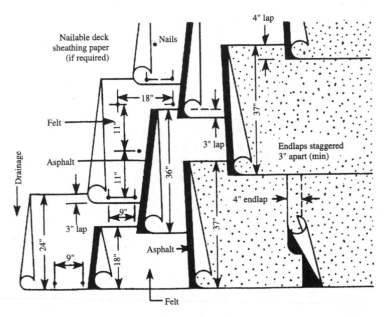

FIGURE 6.3 MBR applied with multiple sheets.

PREPARING THE ROOF

The following are some general guidelines for the care and use of single-ply membrane materials (Figure 6.4).

- Warehouse materials in the manufacturer's original unopened packaging with labels intact.
- Store materials onsite under protective coverings and off the ground.
- Do not store material on the deck in concentrations that impose excessive strain on the deck or structural members.
- Proceed with roofing work only when existing and forecasted weather conditions permit installation in accordance with the manufacturer's recommendations and warranty requirements.
- Take special precautions, as recommended by the manufacturer, when applying roofing at temperatures below 40°F.
- Do not torch-fuse membranes directly to flammable substrates.

✓	Item
	Deliver materials in the manufacturer's original unopened packaging with labels intact.
	Store materials onsite under protective coverings and off the ground.
	Do not store material on the deck in concentrations that impose excessive strain on the deck or structural members.
	Proceed with roofing work only when existing and forecasted weather conditions permit installation in accordance with the manufacturer's recommendations and warranty requirements.
	Take special precautions, as recommended by the manufacturer, when applying roofing at temperatures below 40°F.
	Do not torch-fuse membranes directly to flammable substrates.
	Check with the manufacturer about application requirements.
	Check with the manufacturer about slope requirements.
	The National Roofing Contractors Association (NRCA) recommends that all roofing materials be installed on roofs with positive slope to drainage.
	If UL-classified roofing membranes are required, carefully evaluate the manufacturer's test data to determine compliance with code or other fire-related performance characteristics.
	Ensure that edge nailers, curbs, and penetrations, including drain bases, are in place and properly secured before starting the job so that the roof system can be installed as continuously as possible.
	Ensure that the insulation or base ply is positively attached to the roof deck.

FIGURE 6.4 Preparing the roof.

✓	Item
	Ensure that the bonding agent used to assemble laps and to adhere the single-ply membrane to the roof substrate is permitted by and acceptable to the material supplier.
	Ensure that positive attachment is achieved with partially adhered and mechanically fastened MBR systems that preclude a continuous film.
	Ensure that the roof membrane is applied in such a way that water can run over, or that the membrane has sidelaps and/or endlaps.
	Check that sheets are aligned so that minimum required end- and sidelap widths are maintained.
	Check that membrane laps are watertight. Repair voids and fishmouths within the lap as soon as possible.
	Install temporary water cutoffs at the end of each day's work. Remove them before installing additional insulation or membrane.
	Apply surfacings, when required, in accordance with the membrane supplier's requirements and so that the entire membrane surface is covered.

FIGURE 6.4 *(continued)*   Preparing the roof.

- Check with the manufacturer about application requirements.
- Check with the manufacturer about slope requirements.
- The National Roofing Contractors Association (NRCA) recommends that all roofing materials be installed on roofs with positive slope to drainage.
- If UL-classified roofing membranes are required, carefully evaluate the manufacturer's test data to determine compliance with code or other fire-related performance characteristics.
- Check that edge nailers, curbs, and penetrations, including drain bases, are in place and properly secured before starting the job, so that the roof system can be installed as continuously as possible.
- Make sure that the insulation or base ply is positively attached to the roof deck.
- Find out if the bonding agent used to assemble laps and to adhere the single-ply membrane to the roof substrate is permitted by, and acceptable to, the material supplier.
- Positive attachment is achieved with partially adhered and mechanically fastened MBR systems that preclude a continuous film.
- Ensure that the roof membrane is applied in such a way that water can run over, or that the membrane has sidelaps and/or endlaps.
- Be sure that sheets are aligned so that minimum required end and sidelap widths are maintained.
- Check that membrane laps are watertight. Repair voids and fishmouths within the lap as soon as possible.
- Install temporary water cutoffs at the end of each day's work. Remove them before installing additional insulation or membrane.
- Apply surfacings in accordance with the membrane supplier's requirements, and double check that the entire membrane surface is covered.

Applying Substrate Materials

- Suggested substrate materials for MBR systems include light steel, which does not have the weight of gravel or ballast; primed concrete, which requires torch welding, particularly when not insulated; mechanically-fastened base sheets; and uninsulated assemblies when the deck is wood, gypsum, lightweight insulating concrete, or cementitious wood fiber.

- MBR is not recommended for roofs where there are concentrations of acids, hydrocarbons, or oils. It is not cost effective for wide-open spaces because large rolls of EPDM can be installed less expensively.
- The condition of the substrate is crucial to the performance of the membrane. The surface of the deck must be clean, firm, smooth, and visibly and sufficiently dry for proper application.
- There are different preparation requirements for new construction, replacement, or re-covering over each type of substrate.
- To avoid defects and litigation, the material supplier and the roofing contractor can be responsible only for acceptance of the surface of the substrate that is to receive the roofing system.
- Design considerations and questions about structural soundness are the responsibility of the owner or the owner's representative. Put this in the contract.
- If there is a question about the suitability of the deck and/or its surface, reach an agreement with the material supplier, the roofing contractor, and the owner before proceeding with the work.
- When a primer is required, coat the surface with an amount sufficient to cover the entire surface.
- The amount required varies, depending on the nature of the surface. Allow the primer to dry before the membrane is applied.
- If visual examination indicates deficiencies in primer coverage, apply additional primer for complete coverage and then allow it to dry before applying the membrane.
- The existing roof surface must be suitable to receive a single-ply membrane. Both the NRCA and the RMA recommend putting a re-cover board over existing gravel-surfaced roofs before the membrane system is applied.
- Remove loose and protruding aggregate by spudding, vacuuming, or power brooming to achieve a firm, even surface to receive the re-cover board.
- When the existing roof is smooth or mineral-surfaced, refer to the material supplier's specifications.
- Some specifications permit fully adhering by heat welding, hot mopping, or adhesives.
- Others require base sheets over the existing roof, while still others permit spot mopping, strip attachment, and/or mechanical fastening.

Laying Insulation

- Manufacturing tolerances, dimensional stability, application variables, and the nature of insulation board make it difficult to obtain tightly butted joints.

- Some variance is expected, but the spacing between insulation boards should be the space specified by the material supplier.

- Fill insulation gaps between adhered or mechanically fastened insulation boards according to manufacturer's or architect's instructions.

- Reduce gaps between loosely laid insulation boards by adjusting the boards or adding insulation.

- If the insulation boards appear to be out of square, make a diagonal measurement to confirm the squareness.

- Do not use defective material.

- When composite board, polyisocyanurate foam board, polyurethane foam board, perlitic board, or wood fiberboard insulation is used as the insulation substrate under a torch-applied MBR membrane, install a base sheet as the first layer below the roof membrane. This protects the insulation substrate from the flame and heat of the torch.

- Lap each base sheet a minimum of 2 inches over the preceding sheet; endlaps should be a minimum of 4 inches.

- Adhere the base sheets to the insulation in accordance with the manufacturer's specifications.

Fastening Insulation and Base Plies

- Various mechanical fasteners are used to apply insulation and base plies. They also can be used to attach the single-ply membrane to the roof deck and to attach flashing materials.

- The fasteners often are specified by type, number, and the spacing distance necessary to fulfill the attachment functions.

- Practical considerations often prevent exact spacing, such as 6 inches on center. Reasonable variances from specified spacing distances are expected, but the minimum number of fasteners specified should be used.

- The fastener type and the spacing should be as specified by the material supplier with the understanding that the spacings are average values and the spacing between any two fasteners can vary.

- Should fastener deficiencies be discovered, install additional fasteners as needed and space them appropriately.
- On all nailable roof decks, install a base sheet as the first layer below the roof membrane.
- This base sheet serves as a separating layer between the nailable deck and the roof membrane.
- Lap each base sheet a minimum of 2 inches over the preceding sheet.
- Endlaps should be a minimum of 4 inches.
- Fasten the base sheets according to the manufacturer's recommendations.

Applying Single-Ply Materials

- As stated earlier, there are several attachment methods, depending on the single-ply material.
- System configurations change from specification to specification and are predicated not only on the manufacturer's procedure, but also on site conditions.
- This can, and does, leave some questions about the proper method to use.
- Nonetheless, fundamental procedures for the correct installation of all single-ply membranes remain the same, regardless of the system or specification.
- The strength and waterproofing characteristics of single-ply membrane systems depend on the construction of the laps and the seams between membrane sheets and between the membrane and flashing materials.
- Lap dimensions typically are specified as an amount of overlap between adjacent sheets. Due to the many construction variables already mentioned, some variance can occur. An overlap exceeding the specification is not considered detrimental.
- The material supplier's minimum lap values must be maintained, however, so as not to affect the strength and waterproof integrity of the membrane.
- Fabricate membrane laps so that they are watertight. Repair voids and fishmouths within the lap.
- Frequently, products are supplied with laying lines.
- Use these lines to position the subsequent course of material for compliance with nominal lap requirements.

- Due to the variation in substrates, products, and systems, refer to the material supplier's specifications for minimum requirements.
- If the product does not have laying lines, use the material supplier's stated lap dimension as the minimum lap requirement.
- If examination reveals insufficient lap width, install a strip no narrower than twice the required lap over the deficient lap.
- Use the appropriate attachment method. Check all laps of heat-welded systems for adequate bonding.
- Seal any unbonded area.
- Several factors determine the point at which work begins:
  - For water runoff, begin work at a high point on the roof and move toward the low point. This prevents any drainage water due to overnight rains from working its way under the new roof.
  - Always shingle membrane laps with the flow of water or parallel to the flow of water.
  - Consider the deck type.
  - Metal decks have corrugations that fall in 6 inch modular spacings.
  - As with rigid insulation, fasten the membrane to the top of the flutes to meet Factory Mutual Engineering (FME) specifications.
  - Since the exposed width of most full rolls of exposed membrane is 67½ inches, the edge of the sheet eventually falls between the flutes when the membrane is run parallel to the flutes.
  - When this happens, trim the excess back to the nearest flute top (Figure 6.5a).
  - When the membrane is run perpendicular to the metal deck corrugations, the amount of trimming and membrane waste is less than it would be if the membrane ran parallel to the corrugations (Figure 6.5b).

APPLYING THERMOSET MATERIALS

- Thermoset single-ply material can be applied loose-laid and ballasted, partially adhered, or fully adhered.
- Always roll out the sheet and allow it to relax.

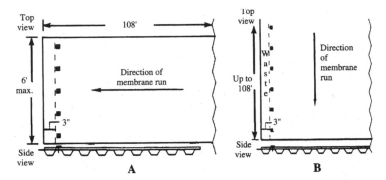

FIGURE 6.5 (a) Membrane running parallel to metal decking might require trimming along entire length of roll. (b) Membrane running perpendicular to the metal deck corrugations. The maximum trim is across the width of the roll.

- Inspect the sheet for defects as it is being rolled out.
- To seal the laps, fold the sheet back onto itself and clean the underside of the top sheet and the upper side of the bottom sheet with the recommended solvent to remove dirt and talc from the lap area.
- Apply adhesive evenly to both surfaces and allow it to set according to the manufacturer's application instructions.
- Then, roll the top sheet over the bottom sheet in a manner that minimizes voids and wrinkles.
- Apply pressure to the lap to ensure contact.

Loose-Laid and Ballasted Method

- The membrane is placed on a substrate without bonding and held in place by ballast.
- The NRCA recommends that consideration be given to mechanically fastening or spot adhering the first layer of insulation to the roof deck under loosely laid roof systems.
- Only enough mechanical fasteners or adhesive need be used to hold the insulation boards in place during installation of the roofing membrane, and to prevent stacking or displacement.

- Attach the second layer of insulation, if used, to the first layer with a compatible adhesive.

- As an alternative procedure for nailable roof decks, both layers can be mechanically fastened to the deck at the same time.

- Adhesive application quantities, mechanical fastener spacing, lap dimensions, and other minimum installation requirements vary from manufacturer to manufacturer.

- Over cellular glass insulation or any other insulation substrate where rough surfaces, protrusions, or sharp edges might affect the service of the membrane, install a separation sheet between the insulation and the roof membrane.

- Fasten the separation sheet only often enough to hold the sheets in place until the roof membrane is applied.

- Roll out and align the sheet so that it overlaps the previous sheet by the required lap width.

- Seal all membrane laps together. Fabricate laps to shed water whenever possible.

- Ballast the loose-laid membrane with rounded stone, such as washed river gravel (Figure 6.6).

- The amount of needed ballast varies with the location and height of the building.

- Substitute concrete pavers for rounded stone ballast if a protective underlayment is installed to protect the roof membrane from the abrasive surface of the paver.

Partially-Attached Systems

- There are three generic ways to partially attach a thermoset (EPDM) membrane (Figure 6.7).

- First method:

  - Roll out the sheet and align it so that it overlaps the previous sheet by the required lap width.

  - Install mechanical fasteners with large washers, or install bars with mechanical fasteners, through the membrane and into the deck.

  - Cover these fasteners or bars with membrane pieces that are bonded to the membrane in a manner similar to the methods used for sealing laps.

  - Then seal all membrane laps together.

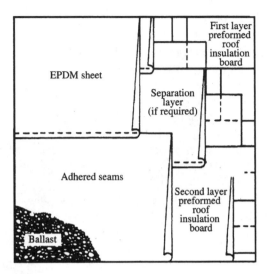

FIGURE 6.6 Loose-laid membrane system.

- Second method:
  - Roll out the sheet and install mechanical fasteners with large washers or bars with mechanical fasteners through the membrane within the area of the lap and into the deck.
  - Seal all membrane laps together so that the fasteners are covered.
- Third method:
  - Mechanically fasten a proprietary fastening system in a predetermined pattern over the separation layer if used.
  - Install the roof membrane over the fasteners.
  - Roll out and align the sheet so that it overlaps the previous sheet by the required lap width.
  - Sandwich the membrane between the parts of the fastening system by installing the second part of the mechanical fastener.
  - Seal all membrane laps together.

Fully-Attached Systems

- For sheet systems, total bonding is usually achieved by applying an adhesive.

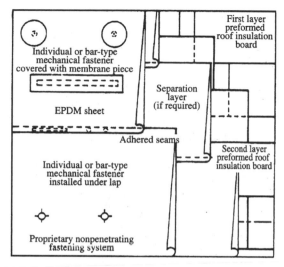

FIGURE 6.7 Partially attached roof system.

- The substrate for adhesive-bonded systems must be smooth, clean, dry, and free from dust, dirt, grease, oil, wax, and loose particles.

- Prime the substrate before applying the adhesive to achieve a greater bond.

- Contact adhesives usually are used in total bonding.

- They are applied to the top surface of the substrate and the bottom of the sheet.

- Some sheets are self-adhering, however, so that the bottom of the sheet adheres to the substrate without an adhesive.

- To apply a fully-attached or total-bonded application, roll out the sheet and align it so that it overlaps the previous sheet by the required lap width.

- Fold the sheet back onto itself and coat the bottom side of the membrane and the topside of the deck with adhesive.

- Avoid getting adhesive on the lap joint area.

- After the adhesive has set according to the adhesive manufacturer's application instructions, roll the membrane into the adhesive in a manner that minimizes the occurrence of voids and wrinkles.

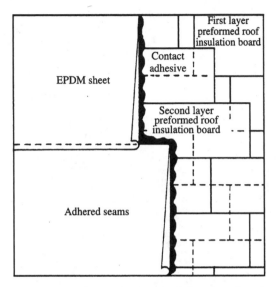

FIGURE 6.8   The mechanics of a fully-adhered roof system.

- Repeat this for the other half of the sheet.
- Take care to adhere the very middle of the sheet (Figure 6.8).
- Then, seal all membrane laps together.
- Fabricate caps to shed water wherever possible.

SEAMING SINGLE-PLY MATERIALS

- A single-ply system is only as strong as its seams.
- There is more than a little truth to this cliché, especially with elastomerics.
- Unlike hot BUR, elastomeric and plastomeric sheets are preformed in the factory.
- Most of the problems with these systems can be traced back to field workmanship, of which lap seaming is a critical part.
- Most contractors and other professionals think that poor workmanship is the most common reason for single-ply roof failure.

- Most rubber-based, single-ply membranes require adhesive for lap seaming.
- A new generation of adhesives and sealants that speed application and improve the performance of rubber-based systems is providing simplified design, ease of inspection, proven performance, and low maintenance.

Bonding Adhesives

- All-purpose bonding adhesives, geared toward roofing applications, adhere EPDM, neoprene, and butyl roof membranes to wood, concrete, metal, and certain insulation surfaces.
- These are contact-type adhesives.
- Their use involves the application of a uniform coat to the back of the rubber membrane, as well as to the surface to which it is to be bonded.
- These successful products have earned wind-test ratings and are accepted by most code authorities.
- Manufacturers advise that the mating surfaces be dry, clean, and free from oil, grease, and other contaminants.
- The best working temperatures are at least 40°F.
- Many EPDM suppliers currently use one-part butyl adhesives to seal their 45 and 60 mil membranes.
- These adhesives normally give consistent peel strengths in the range of 3 to 4½ pounds per inch.
- The application of one-part butyls is somewhat less forgiving than that of other adhesive systems, however.
- The open time or window, defined as the time from when the seam can first be closed to when it no longer grabs, is shorter and varies with weather conditions.
- Keep in mind that one-part butyl requires a heavier adhesive coating. It can skin over and give a false indication that it is dry.
- In addition, it has been found that improperly stirring the adhesive results in inadequate peel strengths.
- Butyl cement begins to cure when exposed to heat and moisture.
- Once a can of adhesive is opened, do not reseal it or use it again because a substantial thickening of the cement occurs in about 48 hours.

- On the other hand, if butyl goes bad, the applicator knows it because the material does not stick.
- The pass/fail criteria for other adhesives are not as obvious.

Splicing Adhesives

- There are lap-splicing adhesives for EPDM, neoprene, and butyl membrane roofing.
- These adhesives provide long-lasting, weather-resistant bonding between the single-ply membranes used in various systems.
- Technology has enhanced the long-term performance of these adhesives.
- The size of the splice determines the application.
- In most cases, for flashing and expansion joint applications, the minimum splice width is 3 inches.
- Contact adhesives must be applied to both of the surfaces being bonded.
- The splice area must be dry and completely free of dust, talc, and other contaminants.
- The best working temperatures are between 40° and 120°F.

Sealants

- Adhesive-seamed, cured, and uncured elastomer membranes all require the use of a lap sealant at the seam edge.
- Most adhesives are vulnerable to moisture until fully cured.
- Some materials take as long as a week to become fully moisture resistant.
- In addition to sealing out moisture and other contaminants, the sealants absorb thermal expansion and resist fatigue, vibration, and attack from microorganisms.
- They are ideal for sealing roofing protrusions, cracks, ductwork, and exterior seams.
- In order to allow the solvents to escape from inside the seam, do not apply the lap sealant for at least two hours.
- If rain is imminent, or if there is a need to leave the jobsite, apply the lap sealant earlier rather than later.

- Repairing minor blistering in the caulked edge is always easier than dealing with ruined seams.
- It is also easy to wander off the membrane edge when you install the lap sealant.
- Due to this, it is important to tool the sealant.
- Ideally, one worker should install the material while a second worker follows behind and tools the edge.
- If one waits too long to tool, the material can harden and start building up on the tool.
- Sometimes a primer is applied to the sheet before the adhesive to enhance seam performance.
- Unlike splice or primer washes, the solvent is applied at full strength.
- The combination of primer and neoprene adhesive usually offers performance similar to that of one-part butyl.
- To apply the adhesive system, first use a splice cleaner on the seam.
- After the area dries, roll the primary butyl adhesive over the entire width of the splice.
- Next, apply a bead of inseam sealant on top of the butyl and roll the seam closed.
- Finally, secure the splice with a lap sealant.
- The inseam sealant is also designed to act as a redundant adhesive.
- If the installer misses a small section of the seam with the butyl, the inseam sealant should still adhere to the EPDM and seal the seam.
- Since the sealant has a greater mass than the butyl adhesive, it is capable of filling any voids or gaps in the seam created by an uneven substrate.
- Be aware that the inseam sealant does not have the green strength to hold the seam together during setup time.
- Therefore, the butyl adhesive must be installed in conjunction with the sealant to avoid fishmouths at the edge of the seam.

Tape

- Butyl-based tape offers the extra mass of adhesive needed when three layers of membrane come together, forming a T-joint.
- According to many roofers, the tape is somewhat more sensitive to dust, dirt, and contaminants than the liquid adhesive.

- Applying the liquid adhesive with a bristle brush can lift contaminants into the adhesive mass.
- On the other hand, tapes have speed and laborsaving advantages, if the mechanic is skilled.
- While many tapes used in the past were made of uncured material, today's tape adhesives are cured.
- Uncured tapes move independently of the roof during expansion and contraction.
- Cured tape is able to move with the sheet.

Heat Welding

- Until recently, one of the easiest ways to tell if an applied single-ply membrane was thermoset or thermoplastic was to look at the seams.
- Thermoset seams were installed with adhesive tape or sealer, while thermoplastic seams were usually heat-welded.
- Several manufacturers are now producing a hot-air weldable EPDM membrane which eliminates the use of adhesives or tapes.
- One product consists of a patented blend of polypropylene and vulcanized rubber.
- The polypropylene gives the membrane heat-welded characteristics and greater oil resistance, while EPDM and carbon black protect the polypropylene from UV degradation.
- During manufacture, a codispersion between the EPDM and polypropylene takes place, with the molecular crosslinking of the polypropylene already in the EPDM formulation.
- Another heat-weldable EPDM is made of polypropylene and a thermoplastic rubber.
- A coextrusion process causes the crosslinking of the two materials, with the top layer containing more vulcanized rubber and the bottom layer more polypropylene.
- The 45-mil membrane is mechanically fastened in seam using 2 inch diameter washers set 18 inches on center.
- A UL Class 90 wind uplift rating and Class B fire rating (smooth surfaced) have already been achieved.
- Tear strength is a consideration in a non-reinforced, mechanically-attached sheet.

- The crosslinking process gives the membrane a 360 psi tear strength across the sheet, which is where wind uplift forces are greatest, according to the manufacturer.
- CSPE, as mentioned earlier in the chapter, is heat-welded.

APPLYING THERMOPLASTIC MATERIALS

PVC, CPE, and PVC blends can be installed by loose-laid, partially attached, or fully adhered methods.

Using the Loose-Laid Method

- On all nailable roof decks, install a separator sheet as the first layer below the roof membrane to serve as a separating layer between the nailable deck and the roof membrane (Figures 6.9a and 6.9b).
- Lap each separator sheet a minimum of 2 inches over the preceding sheet.
- Endlaps should lap a minimum of 4 inches.
- Fasten the separator sheets only as necessary to hold the sheets in place until the roof membrane is applied.
- Roll out the sheet over the separator sheet and inspect the membrane for defects.
- Align the sheet so that it overlaps the previous sheet by the required lap width.
- Seal all membrane laps together using either heat welding or chemical fusion.
- Fabricate laps to shed water wherever possible.
- Apply pressure to the lap to improve the bond.
- Carefully inspect lap edges. Repair unsealed areas, voids, and fishmouths.
- Generally, manufacturers require that exposed membrane edges be finished with a sealant.
- Fasten all membrane perimeters, terminations, and penetrations as required by the manufacturer.
- Ballast the loosely laid membrane with rounded stone, such as washed river gravel.

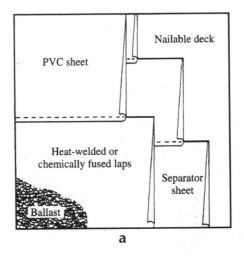

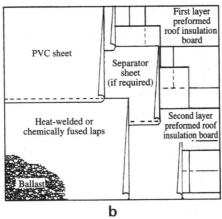

FIGURE 6.9 (a) Loosely laid PVC sheeting over a nailable deck. (b) Loosely laid PVC over an insulated deck.

- The amount of ballast needed varies with the location and height of the building.
- Concrete pavers can be substituted for rounded stone ballast, provided that a protective underlayment is installed to protect the roof membrane from the abrasive surface of the paver.

Using the Partially Attached Method

There are three generic methods for partially attaching thermoplastic membranes (Figure 6.10).

- First method:
  - Mechanically fasten PVC-coated metal discs in a predetermined pattern over the separator sheet.
  - Roll out the membrane and inspect it for defects.
  - Heat-weld or chemically weld the membrane to the PVC-coated metal discs as the membrane is rolled out onto the deck.
  - Align the sheet so that it overlaps the previous sheet by the required lap width.
  - Then seal all membrane laps using heat-welding or chemical fusion.
  - Fabricate laps to shed water wherever possible.
  - Apply pressure to the lap to improve the bond.
- Second method:
  - Install mechanical fasteners with large washers through the membrane and into the deck.
  - Cover these fasteners with membrane patches and heat-weld or chemically fuse the patches to the membrane.
  - Seal the membrane laps using heat welding or chemical fusion.
  - Fabricate the laps to shed water and apply pressure to the lap to improve the bond.
- Third method:
  - Follows the same initial steps.
  - Roll out and inspect the membrane and seal the laps using heat-welding or chemical fusion.
  - Fabricate the laps to shed water and apply pressure to the lap to improve the bond.
  - Then place continuous metal bars in a pre-engineered pattern on top of the membrane and mechanically fasten them to the nailable roof deck.
  - Cover the bars with membrane patches that are heat-welded or chemically fused to the membrane.

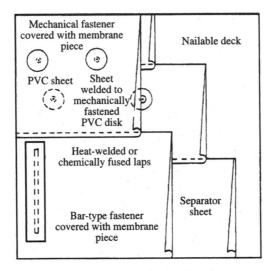

FIGURE 6.10 Partially attached PVC roof system.

Using the Fully Adhering Method

- There are two generic methods for installation of fully adhering thermoplastic membranes (Figure 6.11).
- Solvent-based adhesives:
  - Roll out the sheet and inspect it for defects.
  - Align the sheet so that it overlaps the previous sheet by the required lap width.
  - Fold the sheet back onto itself and coat the bottom side of the membrane and the top side of the deck with adhesive.
  - Avoid getting adhesive on the lap joint area.
  - After the adhesive has set according to the manufacturer's application instructions, roll the membrane into the adhesive in a manner that minimizes the occurrence of voids and wrinkles.
  - Repeat this for the other half of the sheet. Carefully adhere the middle of the sheet.
- Water-based adhesives:
  - Follow the steps for solvent-based adhesives, but avoid getting the adhesive on the lap joint area when you coat the deck.

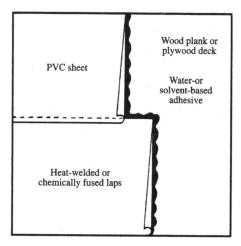

FIGURE 6.11 A fully adhered PVC roof system.

- After the adhesive sets according to the manufacturer's application instructions, roll the membrane into the adhesive in a manner that minimizes the occurrence of voids and wrinkles.

SEAMING THERMOPLASTICS

- PVC, CPE, and the PVC blends are inherently heat-weldable, with typical lap seam strengths of more than 20 pounds per inch.
- While hot-air and solvent-welding procedures are quicker than adhesive seaming, a considerable amount of applicator skill is still required.
- For example, CPE exhibits a higher melting temperature and a narrower welding window than most PVCs.
- It is especially important to conduct weld tests when ambient roof conditions change substantially.
- There are two basic types of welders available—automatic and hand.
- The automatic welder rides on the top sheet, which is being welded to the bottom sheet.

- For this reason, it is possible to start welding in one of two roof corners—the upper right-hand corner or the lower left-hand corner (Figure 6.12).

- To establish these two corners, face the same direction as the run of the membrane. The roof corners to your upper right and lower left are the possible starting points.

- Use the hand welder to weld seams when automatic welding is not possible, such as near a parapet wall or curb detail.

- Hand-welding is also used for many flashing details, applying membrane patches, and repairing poor or unfinished machine-welded seams (Figure 6.13).

- When the gun heats the membrane to a semi molten state, use a hand roller to bond the lap seam.

- A number of interchangeable nozzle tips are designed to match any hand-welding situation.

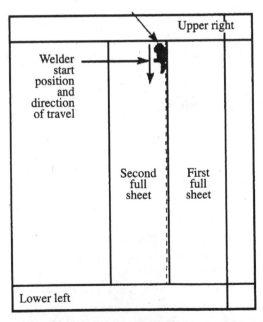

FIGURE 6.12 Welding can begin at one of two points on the roof, the upper right or the lower left.

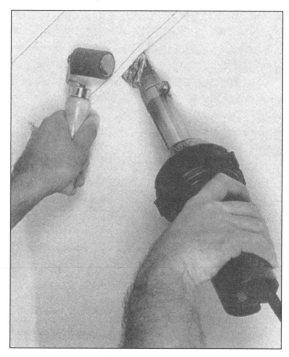

FIGURE 6.13 A hand welder in use.

- For heavily reinforced PVC blends, applying pressure at the right temperature is crucial to a good lap seal.
- Since there are several models and designs, always follow the adjustment procedures outlined in the service manual.

Checking Welded Seams

- Working conditions vary from job to job, from day to day, and even from hour to hour.
- Varying conditions require different heat and speed settings on the welding machine.
- Since conditions vary, it is extremely important that you inspect all welded seams with care.
- Adjust welder speed, temperature, and setup as needed.

- Wait several minutes before testing seam strength, since even tight welds can be separated while they are still hot.

- A good weld is one where the top sheet does not separate from the bottom sheet without damaging the membrane.

- Proper seam width is approximately 1½ inches, but always follow the manufacturer's instructions.

- The welded seam should be smooth and continuously bonded without voids or air pockets.

- The lap edge should be bonded thoroughly.

- Check for gaps in this seam by running a cotter-pin puller along the seam.

- The automatic welder usually misses the first few inches of a seam, so pull back the lap after it has cooled and mark off the area that requires hand welding.

- If there are voids in the weld seam, or if the finished seam is weak, it probably means that the hot-air nozzle did not have enough time to heat the membrane to the necessary molten state.

- To correct this problem, lower the machine's welding speed and make another test run.

- Other problems, such as dirt on the membrane, moisture on the membrane, or variances in the power supply, also can produce voids and poor bonding. Inspect the welded seam and the area around the weld.

- A light browning or burning of the membrane indicates that the nozzle is overheating the weld area.

- To correct this, increase the welding speed slightly and test again.

- If the membrane still burns at the highest speed setting, lower the nozzle temperature by adjusting the heat control dial.

- Brown streaks at the edge of the overlap usually are caused by the outer edge of the nozzle dragging heated dirt and membrane particles across the membrane.

- To prevent this, clean the nozzle with a wire brush.

- Pinch wrinkles can be pulled into the membrane sheets when the automatic welder is not set up properly.

- If pinch wrinkles occur, review the setup and operating procedures given earlier in this chapter and those provided with the welder.

- Repair pinch wrinkles with membrane patches.

APPLYING MODIFIED BITUMEN MATERIALS

MBR can be installed using torch-applied, hot-mopped asphalt or a self-adhered method.

Torch-Applied or Heat-Weld Method

- To apply the membrane, roll out the sheet and inspect it for defects.
- Align the sheet so that it overlays the previous sheet by the required lap width and then re-roll it.
- Heat the underside of the roll to soften the bitumen coating (Figure 6.14).
- Take care not to overheat the top surface of the sheet.
- Most torch-applied membranes come with a factory-installed, burn-off film.
- This film is usually polyethylene, though it can be polypropylene.
- Burn-off films are applied as a release agent and as a compatible material that blends with the MBR coating when heated.

FIGURE 6.14   Making an adjustment on a hand welder.

- This creates an excellent cohesive bond at the lap or seam.
- Films also facilitate bonding to other surfaces.
- Some torch-down products incorporate talc or sand as a release instead of burn-off films.
- To obtain flow of the MBR coatings, apply heat to the burn-off film surface in sufficient amounts to soften and melt the coating.
- When the roll is unwound you should see a flow ahead of the roll and at the sidelap that minimizes voids and wrinkles.
- At the sidelap and all seams, flow should be from a minimum of ¼ inch to a maximum of ½ inch.
- It should be continuous and uninterrupted. No flow at all indicates that insufficient heat was applied.
- A flow more than ½ inch indicates that too much heat was applied.
- Take care to ensure proper lap alignment.
- Membrane laps are sealed together as the sheet is adhered to the deck.
- Fabricate laps to shed water wherever possible.
- Check all seams for proper bonding.
- Check any lap or seam that has no flow for integrity.
- Some manufacturers require that all seams be buttered or troweled.
- In most cases, however, this is not necessary.
- Any installation areas that have an observable flow in excess of ½ inch are not likely to have bonding surfaces.
- This yields the maximum effect because of excess coating flow.
- In some cases, the coating can be almost completely forced away from the bonding surfaces, which reduces maximum bonding potential.
- Excess heat also can affect polyester-reinforced membranes.
- These membranes are susceptible to stretch and thermal breakdown if too much heat is applied.
- Polyester burns and melts, which dramatically reduces performance characteristics.
- Fiberglass-reinforced membranes do not stretch under heat.
- They can, however, experience thermal breakdown of binders, which can result in a loss of performance characteristics.

Hot-Mopped Method

- Roll out and inspect the sheet.
- Then align the sheet so that it overlaps the previous sheet by the required lap width.
- Re-roll the membrane.
- Apply a mopping of hot asphalt immediately in front of the roll.
- Follow the guidelines for heating asphalt as predicated by the equiviscous temperature (EVT) printed on the carton or wrapper, or review Table 5.1.
- Never heat the asphalt to its flash-point temperature.
- Align the membrane plies carefully.
- Stagger or offset endlaps a minimum of 3 feet.
- Apply the first membrane course at the lowest point of the deck. This prevents water from flowing against the lap seams.
- Mop consistently at about 25 pounds per square.
- The mopping action should be continuous and uninterrupted when fully adhered systems are installed.
- Do not feather or taper moppings.
- When applying the hot-asphalt-applied MBR membrane, advance the roll into hot asphalt that has been applied no more than 4 feet in front of the roll.
- Moppings more than 4 feet from the advancing roll can cool and cause false bonding.
- Check all laps and seams for proper bonding.
- Fabricate the laps to shed water whenever possible.
- Flow from the mopping asphalt at laps and seams should be from the torch-applied membrane.
- Use a thermometer to check asphalt temperatures at each point before application.
- For added protection and aesthetics, mineral granules can be sprinkled on the asphalt flow at all laps and seams while the asphalt is still hot.
- Other materials can also be used to surface MBR membranes.
- In addition to their primary function of protecting the membrane from the elements, these surfacings often serve to increase the fire and impact resistance of the roofing system.

- Liquid-applied surfacing materials vary in physical properties and in formulation.
- Surfacing materials and aggregate are applied by various techniques, such as hand spreading and mechanical application.
- They also are applied on a variety of roofs in many climatic conditions.
- These factors and others preclude a high degree of uniformity in applying liquid-applied surfacing materials and aggregate over a roof area.
- Hot-asphalt-applied membranes come with a sand release that is used as an anti-blocking agent and as a surface that facilitates adhesion to the asphalt.
- Hot-asphalt-applied MBR membranes are always SBS modified bitumens with a sanded bonding surface.

Self-Adhered Method

- Cold adhesives are used with some MBR membranes to bond the sheet to the substrate.
- Use the adhesive specified by the material supplier.
- During the application of a fully adhered system, you should observe a continuous, firmly bonding film of adhesive.
- The adhesive should flow out from the membrane to form a seal.
- If aesthetic appearance is a factor, or is dictated by specification, matching granules can be used to cover the flow-out area.
- When using the fully self-adhered application, prime the surface of the roof deck with asphalt primer and allow it to dry.
- Primers are not required if a base sheet is used. Roll out the sheet, inspect it for defects, and align it so that the membrane overlaps the previous sheet by the required lap width.
- Then re-roll it.
- Remove the release paper from the underside of the membrane and roll the membrane over the primed surface in a manner that minimizes voids and wrinkles.
- Take care to ensure proper lap alignment.
- Membrane laps are sealed together as the sheet is adhered to the deck.

- Fabricate laps to shed water whenever possible.
- Apply pressure to both the membrane and lap areas to ensure contact.
- In partially adhered membrane applications, MBR compounds, asphalt, and adhesives are used to secure the roof assembly to the structure.
- These materials are installed in a configuration prescribed by the material supplier, such as spot or strip bonding.
- They often are specified by amount and spacing distances.
- Practical considerations often prevent the application of the exact amount and spacing, such as 12 inches in diameter and 24 inches on center of the bonding agent.
- Reasonable variances are expected.
- The application should result in a firmly bonded membrane.
- The spacing of the bonding agent should be specified by the material supplier.
- The design team and owner should understand that the spacings are average values and that the distance between any two adhesive locations can vary.
- Correct deficiencies by installing an additional bonding agent as needed.
- Space appropriately.
- Determine the scope of the discrepancy and take appropriate remedial action.

FLASHING AT TERMINATIONS

Single-ply membrane terminations at roof edges, parapets, and flashing are treated in a variety of ways depending on the material used.

Cured Elastomeric Membrane Flashing

- Cured elastomeric membrane that is identical to the roof membrane material can be used as flashing material.
- Cut the membrane to size and coat the bottom side of the flashing membrane and the area to be flashed with adhesive.

- After the adhesive has set according to the manufacturer's application instructions, roll the membrane into the adhesive in a manner that minimizes the occurrence of voids and wrinkles.
- Exercise care that the flashing does not bridge where there is a change of direction, i.e., where the parapet wall intersects the roof deck.
- Seal joints where the cured elastomeric membrane flashing material meets with the roof membrane in a manner similar to that used for membrane laps.
- Apply pressure to the laps to ensure contact.

Uncured Elastomeric Membrane Flashing

- Uncured elastomeric membrane can be used in lieu of cured elastomeric material.
- It is especially suited for penetrations, corners, and other flashings where the membrane must be formed or must change direction.
- The uncured membrane is used in a manner similar to cured membrane material.
- Upon curing, however, it permanently shapes to conform to the substrate being flashed.

PVC-Coated or PVC-Clad Metal Flashing

- PVC-coated or PVC-clad metal can be used as flashing material.
- The flat metal stock is cut and fabricated to the proper shape to meet field requirements.
- The sheet metal pieces should not exceed 10 feet in length.
- Fasten the metal in place with fasteners typically spaced 4 to 6 inches on center.
- Round the corners and the edges of the sheet metal flashing to protect the membrane from punctures or chafing.
- Cover the gap with tape.
- Splice the two pieces of metal together with a minimum 5 inch wide piece of PVC membrane.
- Center the piece on the gap so that it extends from the top of the metal flashing out onto the field of the roof.

- Either heat-weld or chemically fuse the PVC membrane to the PVC-clad or PVC-coated metal flashing.

Reinforced PVC Membrane Flashing

- Reinforced PVC membrane flashing can be used in lieu of PVC-coated or PVC-clad metal flashing provided the membrane is first mechanically fastened to the nailable roof deck at the flashing.
- Fasten the reinforced PVC membrane flashing at its upper edge, and shape it to conform to the parapet wall, curb, or edging.
- Then extend the reinforced flashing onto the field of the roof a minimum of 4 inches. Heat-weld or chemically fuse it to the roof membrane.
- Endlaps are to be a minimum of 4 inches.
- Heat-weld or chemically fuse them to the adjoining piece of flashing.
- Various PVC flashing details are shown in Figure 6.15.

MBR Flashing Systems

- MBR flashing materials are applied using a variety of methods, including heat welding, hot asphalt, and adhesive applications.
- The degree of heat required for appropriate heat welding, the variables affecting the application rates and temperatures of any asphalt used, and the amount of adhesive used are subject to many variables, including weather conditions, job conditions, material type, and application method.
- In addition, the material supplier's requirements must be considered.
- Membrane termination follows the same basic procedure at roof edges, parapets, and other flashing applications as the BUR materials described earlier.
- It can be handled in one of three ways.
- Torch-applied MBR membranes can be used as flashing material for masonry, concrete walls, curbs, or other noncombustible substrates.
- Cut the membrane material to size and prime the wall or curb and the area of the roof membrane to be flashed.
- Place the MBR flashing membrane bottom-side up on a piece of plywood or other suitable platform.

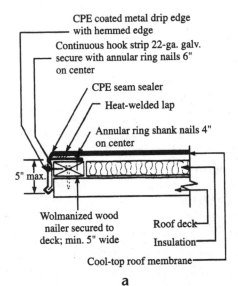

CPE coated metal drip edge with hemmed edge

Continuous hook strip 22-ga. galv. secure with annular ring nails 6" on center

CPE seam sealer

Heat-welded lap

Annular ring shank nails 4" on center

5" max.

Wolmanized wood nailer secured to deck; min. 5" wide

Roof deck

Insulation

Cool-top roof membrane

a

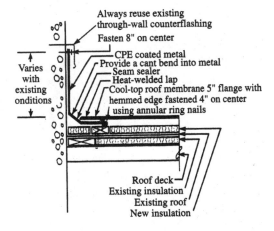

Always reuse existing through-wall counterflashing

Fasten 8" on center

CPE coated metal

Provide a cant bend into metal

Seam sealer

Heat-welded lap

Cool-top roof membrane 5" flange with hemmed edge fastened 4" on center using annular ring nails

Varies with existing conditions

Roof deck

Existing insulation

Existing roof

New insulation

b

FIGURE 6.15 (a) CPE-coated metal drip edge. (b) Installing coated metal flashing under existing wall counter.

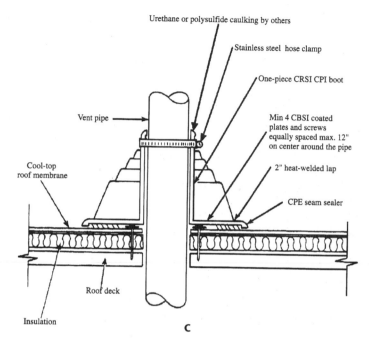

Urethane or polysulfide caulking by others

Stainless steel hose clamp

One-piece CRSI CPI boot

Vent pipe

Min 4 CBSI coated
plates and screws
equally spaced max. 12"
on center around the pipe

Cool-top
roof membrane

2" heat-welded lap

CPE seam sealer

Roof deck

Insulation

c

FIGURE 6.15 *(continued)*   (c) Heated vent-pipe flashing.

- Heat the underside of the membrane with a torch to soften the coating bitumen.
- Set the heated membrane material to the wall or curb in a manner that minimizes the occurrence of voids and wrinkles.
- Take care that the flashing does not bridge where there is a change of direction, for example, where the parapet wall intersects the roof deck.
- Seal the joints where the MBR flashing material meets with the roof membrane in a manner similar to that used for membrane laps.
- Apply pressure to the laps to ensure contact.
- Hot-mopped MBR membranes can also be used as flashing material.
- Follow the directions for torch-applied membranes.

- Instead of torching, apply hot asphalt to the area to be flashed and then to the underside of the membrane.
- Seal the joints where the MBR flashing material meets with the roof membrane in a manner similar to that used for membrane laps.
- Apply pressure to the laps to ensure contact.
- The third method is to use self-adhered MBR membranes as flashing material.
- Cut the membrane material to size, and prime the wall or curb and the area of the roof membrane to be flashed.
- Remove the release paper and set the membrane to the wall or curb in a manner that minimizes the occurrence of voids and wrinkles.
- Take care that the flashing does not bridge where there is a change of direction.
- Seal the joints where the MBR flashing material meets with the roof membrane in a manner similar to that used for membrane laps.
- Apply pressure to the laps to ensure contact.
- Locate the flashing terminations a minimum of 8 inches above the level on the roof.
- Counterflashing should follow the material supplier's specifications.
- Some material suppliers have a maximum height limitation for flashings.
- Refer to the material supplier's recommendations.
- A cant strip can be installed to modify the angle between the roof deck and the vertical surface.
- For heat-welded applications, make cant strips from a flame-resistant material and cover them with a base felt.
- Consult the manufacturer's instructions for recommendations regarding the use of cant strips.
- Install the membrane before the flashing is applied and above the plane of the roof, above the cant, and up the vertical surface as specified by the material supplier.
- Do not run the roof membrane up the vertical surface to act solely as a base flashing.
- Remove and reinstall areas of loose, inadequately or improperly bonded flashings using the same materials as in the original application.
- Cut out, rebond, and reinstall laps that are inadequately bonded.

- Inspect flashing seams with care. Repair unsealed areas, voids, and fishmouths. Smoothly finish exposed edges or torch-applied MBR membrane with a hot, rounded trowel.

SURFACING SINGLE-PLY ROOFS

- Some MBR membranes have factory-applied surfacings that provide weather resistance or improve the appearance of the roof membrane.
- Some membranes require that the surfacings or coatings that provide these properties be applied in the field.
- The following is a description of the more common surfacing options for MBR systems.

Ballast

- Loosely-laid MBR systems require the use of ballast to build resistance to wind-uplift forces.
- The ballast is usually rounded stone, such as washed river gravel.
- The amount of ballast needed varies with the location and height of the building.
- Concrete pavers can be substituted for rounded stone ballast, provided that the roof membrane is protected from the abrasive surface of the paver.

Aggregate

- For fire resistance, the application of a flood coat and aggregate surfacing can be required.
- The aggregate should be essentially opaque and of a size specified by the manufacturer or the architect.
- Embed the aggregate in a flood coat of hot asphalt.

Mineral Granules

- For fire resistance, some membranes require the field application of mineral granule surfacing.

- For each roof square, embed approximately 50 pounds of No. 11 roofing granules in an emulsion coating that has been applied to the membrane at the approximate rate of 2 gallons per roof square (see manufacturer's instructions).

Emulsion Coatings

- For aesthetics, some membranes require the application of an asphalt-emulsion top coating.
- Application rates and techniques vary from manufacturer to manufacturer.

Reflective Coatings

- Fibered and non-fibered aluminum/asphalt coatings can be applied to most smooth-surfaced MBR membranes.
- Application rates and techniques vary depending on the manufacturer.

Factory-Applied/Self-Surfaced Membranes

- Some MBR membranes might not require the field application of additional surfacing materials.
- These include membranes with factory-applied mineral granules or metal foils and some smooth-surfaced systems.

ASPHALT ROOFS

Asphalt products are the most commonly used of all roofing materials. They are specified on low-sloped and steep-pitched roofs and are widely used for covering homes and for commercial projects.

- Asphalt roofing can be classified into three groups:
  - Roll roofing
  - Shingles
  - Underlayment

- Roll roofing and shingles are outer roof coverings. They are exposed to the weather and designed to withstand the elements.

- Saturated felts are inner roof coverings. They provide the necessary underlayment protection for the exposed roofing materials.

- Asphalt flux:
  - Asphalt flux is used to manufacture asphalt roll roofing products and asphalt shingles.

  - It is obtained from the fractional distillation of petroleum that occurs toward the end of the refining process.

  - Asphalt flux is sometimes processed at the oil refinery and delivered to the asphalt material manufacturer in a state that satisfies the manufacturer's specifications.

- Many manufacturers, however, purchase the flux and do their own refining.
- Organic felts:
  - Organic felts and glass-fiber base mats are the most common reinforcement in asphalt roofing materials.
  - Organic felts are produced from various combinations of rag, wood, and other cellulose fibers.
  - Glass-fiber base mats are composed of inorganic, continuous or random, thin glass fibers that are bonded firmly together with plastic binders.
  - Generally, the glass fibers in these mats are reinforced with additional chopped glass-fiber strands or with continuous, random, or parallel glass-fiber filaments.
- Surfacing of asphalt products:
  - Fine mineral granules are dusted on the surfaces of smooth asphalt roll roofing materials and on the back side of mineal surfaced asphalt roll roofing materials to prevent the rolled surfaces of the roll from sticking together after the material is wound into rolls.
  - Finely ground minerals are also dusted on the back side of asphalt shingles to prevent the shingles from sticking together in the package.
  - Talc and mica are the most frequently used mineral surfacings.
  - They are not intended to be a permanent part of the finished product and gradually depart from exposed surfaces after the roofing materials are applied.
  - Coarse mineral granules are used on some asphalt roll roofing products and on asphalt shingles to protect the underlying asphalt coating from the impact of light rays.
  - The granules are opaque, dense, and graded for maximum coverage.
  - Coarse mineral granules also increase the fire-resistance rating of the asphalt roofing product.
  - Coarse surfacing is available in a wide range of colors and color blends, which improves the adaptability of asphalt roofing materials to different types of buildings.
  - The range of colors also provides a large variety of design possibilities.

- The materials most frequently used for coarse mineral surfacing are either naturally colored slate, naturally colored rock granules, or rock granules colored by a ceramic process.

- The ceramic-coated mineral granules are applied to the top surface of the shingles during the manufacturing process.

- This provides the widest range of colors available in any type of roofing material. They range from white and black to hues of red, brown, and green.

USING ASPHALT SHINGLES

Asphalt shingles are the most common roofing material used today. They are manufactured as strip shingles, interlocking shingles, and large individual shingles in a variety of colors and weights.

Types of Asphalt Shingles

- Strip shingles:
  - They are rectangular.
  - They measure approximately 12 inches wide and 36 inches long.
  - They have as many as five cutouts along the long dimension.

fastfacts

➤ *Most strip shingles are available with strips or dabs of a factory-applied, self-sealing adhesive which is a thermoplastic material that is activated by the sun's heat after the shingle is installed.*

➤ *Exposure to the sun's heat bonds each shingle securely to the one below for greater wind resistance.*

➤ *This self-sealing action takes place within a few days during the spring, summer, and fall.*

➤ *In winter, the self-sealing action varies depending on geographical location, roof slope, and the orientation of the house on the site.*

- Cutouts separate the shingle's tabs, which are exposed to the weather and give the roof the appearance of being made up of a larger number of individual units.

- Strip shingles are also manufactured without cutouts to produce a much different effect.

- The three-tab shingle is the most common strip shingle.

- Strip-shingle tabs can be trimmed or offset to obtain straight or staggered butt lines.

- Strip shingles can also be embossed or built up from a number of laminations of base material to give a three-dimensional effect.

- Each of these shingle characteristics—staggered butt lines, embossing, and laminations—can be combined in various ways to create textures on the finished roof surface that resemble wood, slate, or tile.

- Fiberglass laminated shingles carry a Class A fire rating from Underwriters' Laboratories, Inc. (UL).

- Single-ply asphalt shingles usually have Class C fire ratings.

- Fiberglass also offers greater resistance to warping, rotting, blistering, and curling. It contains more coating asphalt to better protect the structure from weather elements.

- Interlocking shingles:

  - They provide resistance to strong winds.

  - The shingles come in various shapes and with different locking devices that provide a mechanical interlock.

fastfacts

➤ *In recent years, the laminated two-ply asphalt shingle has become very popular.*

➤ *Because two layers of shingles are laminated together, there is more weight and better protection than with a one-ply, three-tab shingle.*

➤ *Most three-tab shingles have a 15 to 20 year limited warranty compared to 25 to 30 years for laminated shingles. That is 5 to 15 years more service.*

- Large individual shingles are generally rectangular or hexagonal in shape.

USING ROLL ASPHALT ROOFING MATERIALS

Roll roofing is sometimes used for low-cost housing and utility buildings, such as sheds, garages, and barns because it is inexpensive.

- Roll roofing is an asphalted material that is manufactured and surfaced like shingles.

- It comes in rolls that are 36 to 38 feet long and approximately 36 inches wide.

- Some mineral-surfaced roll roofings are manufactured with a granule free selvage edge that indicates the amount by which each succeeding course should overlap the preceding course.

- Follow the manufacturer's recommendations with respect to toplap, side-lap, and end-lap.

- The amount of overlap determines how much of the material is exposed to the weather and the extent of roof surface coverage.

- Overlap also determines if a single or a double layer of roll roofing covers most of the surface.

- Roll roofing is also a common flashing material.

fastfacts

➤ *Most manufacturers certify asphalt shingles to wind-performance test standards on a continual basis through independent third-party testing laboratories.*

➤ *In this standard test, which has been used for many years, shingles are applied to a roof deck according to the manufacturer's specifications, sealed under controlled conditions, and then subjected to ongoing wind conditions for 2 hours.*

➤ *The asphalt shingles are monitored continuously throughout the test.*

USING SATURATED FELT UNDERLAYMENTS

This type of asphalt roofing product consists of a dry felt impregnated with an asphalt saturant. It is used primarily as an underlayment for asphalt shingles, roll roofing, and other types of roofing materials, and as sheathing paper.

• Saturated felts are manufactured in different weights for use as regular or heavy-duty underlayment.

PREPARING FOR ASPHALT ROOFING MATERIALS

Apply asphalt roofing materials over a solid-sheathed deck surface.

• If plywood is used as the deck material, make sure that it is exterior plywood of the correct minimum thickness:
  • It should be $\frac{15}{32}$ inch thick for 16 inch rafter separation.
  • Or $\frac{5}{8}$ inch thick for 24 inch rafter separation.
  • Separate the plywood panels by at least $\frac{1}{16}$ inch to allow for expansion.
• Prior to the application of asphalt roofing materials on plywood or wooden planks, inspect the deck for delaminated areas on the plywood, warped boards, loose knots or improper nailing.
• Check for excessively resinous areas. Wood resin can chemically affect asphalt roofing materials.
• The thickness, type, grade, and installation of other deck materials should conform to local building code regulations.

Using Nails

✔ Use large-headed, sharp-pointed, hot-dipped, and galvanized nails.
✔ Steel or aluminum nails with barbed or otherwise deformed shanks can also be used.
✔ The best roofing nails are made from 11 or 12 gauge wire.
✔ They have heads that are $\frac{3}{8}$ or $\frac{7}{16}$ inch in diameter.

fastfacts

➤ *The underlying structure should provide a rigid deck surface that does not sag, shift, or deflect under the weight of the roofing, the workers installing the materials, or the snow loads the roof might have to support.*

➤ *An unstable structure can lead to movements that affect the integrity of the roofing materials and flashings.*

➤ *Warping, as well as other problems, can result even with well-seasoned lumber if the attic space under the roof deck or the spaces between the rafters and beams are improperly ventilated in a cathedral ceiling or mansard roof.*

➤ *Poor ventilation can result in an accumulation of moisture that eventually condenses on the underside of the roof deck.*

➤ *Asphalt roofing materials can be used on cement roof decks, but the attachment should be applied strictly according to manufacturer's recommendations.*

- Make sure that all nails are long enough to penetrate the roofing materials.
- The nails should extend through plywood decks and at least ¾ inch into wooden plank decks.
- If a fastener does not penetrate the deck properly, remove the fastener and repair the hole in the shingle with asphalt plastic cement or replace the entire shingle. Then place another fastener nearby.
- For asphalt products, use nails of appropriate lengths as indicated in Table 7.1, unless other specifications are given by the architect or manufacturer or other authorities.
- Do not nail into or above factory-applied adhesives.
- Carefully align each shingle.
- Whenever possible, make sure that no cutout or end joint is less than 2 inches from a nail in an underlaying course.
- To prevent buckling, start nailing from the end nearest the shingle just laid and proceed across.

TABLE 7.1 Recommended Nail Sizes

NAIL SIZES	
PURPOSE	NAIL LENGTH IN INCHES
Roll roofing materials on new decks	1
Strip or individual shingles on new decks	1-1/4
Re-roofing over old asphalt shingles	1-1/4 - 1-1/2
Re-roofing over old wooden shingles	1-3/4
Re-roofing over 300-pound or heavier asphalt shingles	1-1/2 - 1-1/4

- To prevent distortion, do not attempt to realign a shingle by shifting the free end after two nails are in place.
- Place the fasteners according to the shingle manufacturer's specifications.
- Align the shingles properly to avoid exposing fasteners in the course below.
- Drive the nails straight and stop them flush with the shingle surface.
- Do not sink the nail into or break the shingle's surface.

USING STAPLES

Staples can be used in place of roofing nails on a one-for-one basis. If there is any doubt about the acceptability of staples in a particular application, consult the shingle manufacturer.

- Roofing staples have a 1 inch crown, legs long enough to penetrate ¾ inch into sheathing, and chisel points.
- Pneumatic guns and special hammers can be adjusted for staples with leg lengths up to 1½ inches.
- Some types of guns have a guide that helps the roofer apply shingles with uniform exposure.
- Staples can be used if the roof slope is 4 in 12 or greater.
- At least six staples per strip are required.

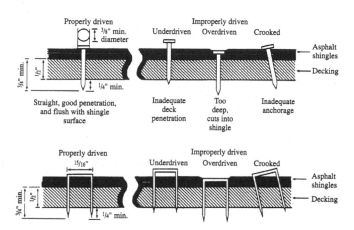

FIGURE 7.1 Good and bad nailing and stapling positions.

- At the eaves, rakes, hips, and ridges, and where shingles touch flashing, use nails instead of staples because of their greater holding power.

- Drive staples parallel to the length of the shingle with properly adjusted pneumatic equipment in order to obtain the required penetration into the deck.

- Figure 7.1 illustrates examples of workman-like nailing and stapling.

FOLLOWING MANUFACTURER'S INSTRUCTIONS

Gypsum concrete plank and tile, fiberboard, or similar non-wood deck materials require special fasteners and/or details to adequately anchor asphalt roofing materials. In such cases, it is recommended that the deck manufacturer's fastening specifications be followed in order to assure that the responsibility for the deck material's performance is placed upon the deck material manufacturer.

SPECIFYING UNDERLAYMENTS

- When a single-layer underlayment is required, apply one layer of No. 15 asphalt-saturated, non-perforated felt horizontally.

- Heavier underlayment is not necessary, but might be required by codes in some areas.

- Lap all felt sheets a minimum of 2 inches over the preceding felt sheet.

- End-laps should be a minimum of 4 inches.

- Nail the felts under the lap only as necessary to hold the felts in place until the asphalt roofing material is applied.

- Laps can be sealed with plastic asphalt cement as required.

- If specifications call for a double-layer underlayment:

  - Apply two layers of No. 15 (minimum) asphalt-saturated, non-perforated felt horizontally.

  - Apply a 19 inch wide starter sheet to the eaves.

  - Cover the starter sheet with a full-width sheet.

  - Lap succeeding sheets 19 inches over the preceding sheets, with a 17 inch exposure.

  - End-laps should be a minimum of 6 inches.

  - Backnail the felts under the laps only as necessary to hold the felts in place until the asphalt roofing material is applied.

  - Laps can be sealed with plastic asphalt cement as required.

- In locations where the January mean temperature is 30°F or lower, apply two plies of No. 15 felt or one ply of No. 50 felt.

- Set this in hot asphalt or mastic regardless of the slope, or in an adhered bitumen membrane underlayment to roof decks with slopes less than 4 inches per foot.

- Work from the eaves to a point 24 inches inside the building's inside wall line to serve as an ice shield.

- The limitations imposed by the slope, or pitch, of the roof deck are factors that must be considered when the roof is designed and underlayments specified.

- For example, self-sealing strip shingles with tabs can be applied on roof decks that have a slope of 4 inches per 12 inches or more if at least one layer of No. 15 asphalt-saturated, non-perforated felt is applied horizontally to serve as the underlayment.

- Self-sealing strip shingles, with tabs, can be applied to roof decks with a slope of 3 inches per 12 inches or more if at least two layers of No. 15 asphalt-saturated, non-perforated felt are applied horizontally to serve as the underlayment.

- Roof decks with a slope of 2½ inches in 12 inches or more must have at least two layers of No. 15 asphalt-saturated felt set in hot asphalt or mastic to serve as the underlayment before self-sealing strip shingles, with tabs, can be installed.

- Laminated asphalt shingles, individual lock-down shingles, and self-sealing strip shingles without tabs can be applied on roof decks with a slope of 4 inches per foot or more if at least one layer of No. 15 asphalt-saturated, non-perforated felt is applied horizontally to serve as the underlayment.

FASTENING ASPHALT ROOFING MATERIALS

When using roofing nails to fasten shingles, drive the nail shaft into the center of all four fastener locations. When using staples, drive the staple crown into the center of both inside locations. Apply the outermost staple leg on center at both end locations.

ATTACHING THREE-TAB STRIP SHINGLES

- Fasten shingles (both the conventional 36 inch long shingles and the metric shingles) with four fasteners in areas with normal weather conditions.

- When the shingles are applied with a 5 inch exposure (5⅝ inches for metric), apply the fasteners on a line ⅝ inch above the top of the cutouts, 1 inch in from each end, and centered over each cutout.

- For high-wind areas, use six fasteners instead of four.

- The fastener locations should be on a line ⅝ inch above the top of the cutouts, 1 inch in from each end, and 1 inch to the left and right of center of each cutout.

ATTACHING NO-CUTOUT STRIP SHINGLES

- Fasten shingles (both the conventional 36 inch long shingles and the metric shingles) with four fasteners.

- When the shingles are applied with a 5 inch exposure (5⅝ inches for metric), the fastener locations should be on a line 5⅜ inches

(6⅛ inches for metric) above the butt edge and located 1 inch and 12 inches (1 inch and 13 inches for metric) from each end.

- For high-wind areas, fasten no-cutout strip shingles with 6 fasteners.
- When the shingles are applied with an exposure of 5 inches (5⅝ inches for metric), the fastener locations should be on a line 5⅝ inches (6⅛ inches for metric) above the butt edge and located 1 inch, 11 inches, and 13 inches (1 inch, 12 inches, and 14 inches for metric) from each end.

INSTALLING ASPHALT SHINGLES

- If a roof surface is broken by a dormer or valley, start applying the shingles from a rake edge of the roof and work toward the dormer or valley.
- If the roof surface is unbroken, start at the rake that is most visible.
- If both rakes are equally visible, start at the center and work both ways.
- On hip roofs, start at the center and work out toward the edges in both directions.
- No matter where the application begins, apply the shingles across and diagonally up the roof. This ensures that each shingle is fastened properly.

APPLYING THE STARTER COURSE

- The starter course can be a 7 inch or wider strip of mineral-surfaced asphalt roll roofing material or a row of shingles trimmed to the manufacturer's specifications.
- The starter course or strip protects the roof by filling in the spaces under the cutouts and joints of the first course of shingles.
- It should overhang the eaves and rake edges by ¼ to ¾ inch.
- If self-sealing shingles are used for the starter strip, remove the tab portion of each shingle and position the remaining strip with the factory-applied adhesive face up along the eaves.
- Trim at least 3 inches from the end of the first shingle in the starter strip. This ensures that the cutouts of the first course of shingles are not placed over the starter strip joints.

fastfacts

➤ *Straight-up application or racking can result in less than the recommended number of nails being used because of the manner in which the shingles are applied.*

➤ *Racking requires that part of the shingles in some courses be placed under those already applied in the course above. Because part of the shingle is hidden, it can be overlooked when the shingle is fastened.*

➤ *When tiles are installed diagonally, each shingle is completely visible until it is covered by the course above.*

➤ *Racking can also accentuate shading tendencies.*

- Fasten starter strips parallel to the eaves along a line 3 to 4 inches above the eaves.

- Position the fasteners so that they are not exposed under the cut-outs in the first course (Figure 7.2).

- If shingles without a self-sealing adhesive are used for the starter strip, remove the tab portion of each shingle and position the remaining strip along the eaves.

- Complete the procedure using the instructions above.

- If roll roofing is used for the starter strip, nail along a line 3 to 4 inches above the eaves.

- Space the nails 12 inches apart.

- If more than one piece of roll roofing must be used, lap the end joint 2 inches and cement it.

APPLYING FIRST AND SUCCESSIVE COURSES

- The first course is the most crucial. Be sure it is laid perfectly straight.

- Check the first course against a horizontal chalkline regularly during application.

- A few vertical chalklines aligned with the ends of the shingles in the first course ensure the proper alignment of the cutouts.

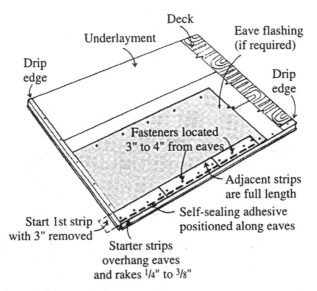

FIGURE 7.2 Starter course.

- If applying three-tab shingles or roll roofing for the starter strip, bond the tabs of each shingle in the first course to the starter strip by placing a spot of asphalt plastic cement about the size of a quarter on the starter strip beneath each tab.

- Then press the tabs firmly into the cement. Avoid excessive use of cement, as this can cause blistering.

CREATING THREE-TAB STRIP SHINGLE ARCHITECTURAL PATTERNS

There are several different methods of applying 3-tab strip shingles.

Six Inch Pattern

- This design is often called "the cutouts that break on joint halves."
- The 6 inch pattern starts each succeeding course with shingles that are 6 inches smaller than the starter shingle in the preceding course.

fastfacts

➤ *The designs usually correspond to the amount removed from the first shingle in each successive course.*

➤ *Removing different amounts from the first shingle staggers the cutouts so that one course of shingles does not line up directly with those of the course below.*

➤ *Start first courses of shingles with full or cut shingles, depending on the pattern desired.*

- The first course starts with a full-length shingle.

- The second course starts with a shingle that has had 6 inches removed.

- The third course starts with a shingle that has had 12 inches removed, and so on through the sixth course, which starts with a shingle that has had 30 inches removed.

- Adjacent shingles in each course are all full length.

- The seventh course again starts with a full-length shingle, and the pattern is repeated every six courses. Figure 7.3 illustrates the pattern.

Five Inch Pattern

- This application is similar to the 6 inch pattern, except that the second through seventh courses are each 5 inches smaller than the preceding one.

- The first course begins with a full shingle.

- The second course starts with 5 inches removed from the first shingle, and so on through the seventh course, which has 30 inches removed from the first shingle.

- Adjacent shingles in each course are full length.

- The pattern is repeated every seven courses.

- One alternative method calls for starting the eighth course with 11 inches removed from the first shingle (Figure 7.4).

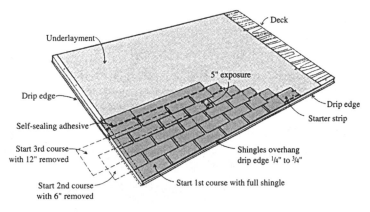

FIGURE 7.3 Six inch pattern.

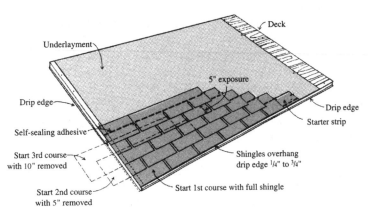

FIGURE 7.4 Five inch pattern.

Four Inch Pattern

- This one is often called "the cutouts that break on joint thirds."
- The four inch pattern design is illustrated in Figure 7.5.
- Start the first course with a full shingle.
- Start the second course with 4 inches removed from the first shingle, the third course with 8 inches removed, and so on through the ninth course, which has 32 inches removed from the first shingle.
- Adjacent shingles in each course are full length.

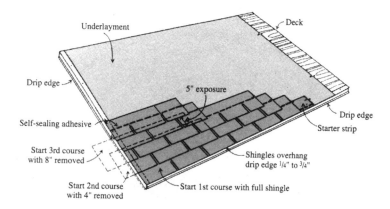

FIGURE 7.5 Four inch pattern

- The tenth course begins with a full-length shingle again, and the pattern is repeated every nine courses.

Spacing

- This pattern can be achieved by removing different amounts of material from the rake tab of succeeding courses in accordance with the following principles.
- The width of any rake tab should be at least 3 inches.
- The cutout centerlines of any course should be located at least 3 inches laterally from the cutout centerlines in both the courses above and the courses below.
- The rake tab widths should not repeat closely enough to cause the eye to follow a cutout alignment.
- Starting the first course with a full-length strip indicates the length of the starting tab for each succeeding course. This procedure is necessary to produce satisfactory random spacing.

Ribbon Courses

- The use of a ribbon course every fifth course strengthens the horizontal rooflines and adds a distinctive, massive appearance that many homeowners find desirable.

fastfacts

➤ *Never use an alignment system where shingle joints are closer than 4 inches to one another.*

➤ *Offset cutouts between the first two layers as explained for six inch, four inch, or random spacing patterns.*

➤ *Very steep slopes tend to reduce the effectiveness of factory applied, self-sealing strip adhesives, especially on the colder or shaded areas of the roof.*

- A preferred method of ribbon course application involves a special starting procedure that is repeated every fifth course.
  - From the tip of a 12 inch wide strip shingle, cut away 4 inches.
  - This provides an unbroken strip 4 x 36 inches in size and a strip 8 x 36 inches in size that contains the cutouts.
  - Lay the 4 x 36 inch strip along the eaves.
  - Cover this strip with the 8 x 36 inch strip, with the bottom of the cutouts laid down to the eaves.
  - Lay this first course of full shingles (12 x 36 inches) over the other layers (see Figure 7.6) with the bottom of the cutouts laid down to the eaves.
  - Repeat this procedure every fifth course, or after a predetermined number of courses, by aligning the ribbon course with the tops of the cutouts of the fifth course.
  - This method produces a triple-thickness buttline every fifth course.

Applying Flashing

Flashing controls moisture entry at ridges, hips, and valleys; around pipes that protrude through the roof's surface; at chimneys and dormers; and wherever a roof butts against a wall. Careful attention to flashing is essential to good roof performance, regardless of the type of roof construction or its cost.

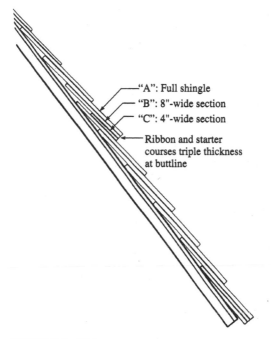

"A": Full shingle
"B": 8"-wide section
"C": 4"-wide section
Ribbon and starter courses triple thickness at buttline

FIGURE 7.6 Ribbon courses.

- Several materials can be used for flashing:
  - Roll roofing is the most widely used, and is generally acceptable at ridges, hips, and valleys.
  - The best flashing for any purpose, however, is corrosion-resistant metal. The metal can be 0.019 inch thick aluminum, 26 gauge galvanized steel, or 16 ounce copper.
  - Vinyl flashing and plastic sleeves around pipes are also used.
- Long lasting roofs require maintenance.
- An important part of maintenance is to make certain that flashings are holding up. Make upkeep a part of your yearly contracted maintenance business.
- When using metal flashing, remember that metals react to one another in the presence of water.
- This reaction, called galvanic action, causes one of the metals to corrode.

- Metals are ranked in the order in which they react to other metals when they are in direct contact.
- This ranking is called the electromotive series.

Common Construction Metals

Metals most often used in construction are aluminum, zinc, steel, tin, lead, brass, copper, and bronze.

- If aluminum flashing is applied with steel nails, the aluminum corrodes because it is higher than steel in the electromotive series.
- If copper flashing is applied with steel nails, the steel corrodes.
- The farther apart the metals are in the series, the sooner and faster corrosion occurs.
- The simplest way to avoid galvanic action is to attach metals with fasteners of the same metal.
- One alternative is to use a nonmetallic washer such as neoprene or asphalt, which keeps the metals from touching each other.
- However, the washers can degrade and must be maintained.

BUILDING VALLEYS

- Construct valley flashings only after the deck has been properly prepared by applying a saturated felt underlayment, as shown in Figure 7.7.
- Regardless of the type of flashing used, the valley must be smooth, unobstructed, of sufficient capacity to carry water away rapidly, and capable of withstanding occasional water backup.
- The construction procedures for building different types of valleys with various materials follow. As might be expected, several steps are the same for each type of valley. Unless indicated otherwise, follow these general rules for each application method:
  - Use only enough nails to hold strips of felt or material smoothly in place.
  - Nail the material along one side of the valley first.
  - Place nails in a row 1 inch from the edge of the strip.
  - As the nailing proceeds along the second side, press the roofing material firmly in place in the valley.

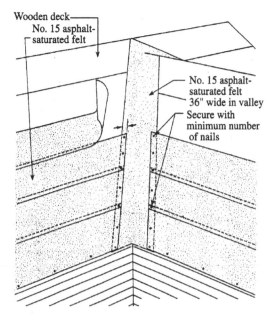

FIGURE 7.7 Valley flashing; felt underlayment.

- Before the roofing material is applied, snap two chalklines along the full length of the valley.
- Place one chalkline on each side, so that the chalklines are 6 inches apart at the ridge, i.e., 3 inches from the center of the valley along each of the intersecting roofs.
- The chalklines should diverge at the rate of ⅛ inch per foot as they approach the eaves.
- For example, a valley 8 feet long is 7 inches wide at the eaves, and a valley 16 feet long is 8 inches wide at the eaves.
- The chalklines serve as a guide by which to trim the last unit in order to ensure a clean, sharp edge.

Open-Valley Using Roll Roofing Material

- The application of mineral-surfaced asphalt roll roofing material, colored to match or contrast with the roof covering, is satisfactory for open-valley flashing.

- This type of application is illustrated in Figure 7.8.

- Center an 18 inch wide layer of mineral-surfaced asphalt roofing material in the valley with the mineral-surfaced side down.

- Cut the lower edge to conform to and be flush with the eave flashing strip.

- When the flashing material must be lapped, lap the ends of the upper piece over the lower piece by at least 12 inches and secure it with asphalt plastic cement.

- On top of the first strip of material, center another strip of 36 inch wide material with the mineral-surfaced side up.

- Secure this strip with nails in the same manner as the underlying 18 inch strip of material.

- Follow the general instructions for holding the material in place.

- Align the material along the chalklines and clip the upper corner of each end to prevent water from penetrating under the courses.

- Cement the roofing material to the valley lining with a spot of plastic asphalt cement.

Open Valley Using Metal Flashing

- Open valleys can be formed by laying strips of at least 26 gauge galvanized metal, or an equivalent non-corrosive, non-staining

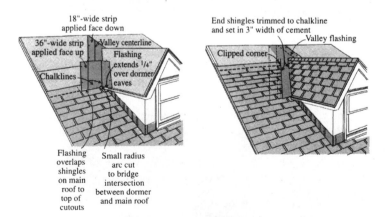

FIGURE 7.8 Open valley using mineral-surfaced asphalt roll roofing.

material, in the valley angle and lapping the roofing material over the metal on either side.

- This leaves a space between the edges of the roofing material to channel water down the valley.

- Extend valley flashing metal at least 9 inches from the centerline of the valley in each direction at the ridge and make it wide enough at the eaves to extend at least 4 inches under the roofing material.

- Form a splash diverter rib not less than 1 inch high at the flow-line from the metal. Flashing metal pieces should not be longer than 10 feet.

- Install metal flashing at open-valley locations as follows. Center a 36 inch wide layer of No. 15 asphalt-saturated, non-perforated felt in the valley over the previously applied underlayment.

- Cut the lower edge to conform with and to be flush with the eaves flashing strip.

- When this sheet must be lapped, lap the ends of the upper piece over the lower piece by at least 12 inches and secure it with plastic asphalt cement.

- Follow the general directions for holding the materials in place.

- Slit the center of the portion attached to the main roof and nail it into place. Then apply the main roof courses to cover the portion of the last ridge shingle on the main roof. Snap a chalkline so that the shingles on the main roof continue the same alignment pattern on both sides of the dormer.

Woven Valleys

- The woven valley treatment is limited to strip-type shingles.

- Individual shingles cannot be used because nails might be required at or near the center of the valley lining.

- In order to avoid placing a nail in an overlapped shingle too close to the center of the valley, it might be necessary to cut short a strip that would otherwise end near the center of the valley and then continue from this cut end over the valley with a full-length strip.

- This method has the advantage of increasing the coverage of the shingles throughout the length of the valley.

- The method also adds to the weather resistance of the roof in an area that might otherwise be vulnerable.

- To install flashing at woven valley locations, center a 36 inch wide strip of No. 15 asphalt-saturated, non-perforated felt in the valley over the existing No. 15 asphalt-saturated felt underlayment.
- Use only enough nails to hold the sheet smoothly in place.
- Lay the first course of material along the eaves of one roof area up to and over the valley.
- Extend it along the adjoining roof area for at least 12 inches.
- Then lay the first course along the eaves of the intersecting roof area and extend it over the valley on top of the previously applied shingle.
- Lay successive courses alternately, first along one roof area and then along the other, weaving the valley shingles over each other, as shown in Figure 7.9.
- Press the shingles into the valley tightly and nail them in place.
- Locate no nails closer than 6 inches to the valley's centerline.
- Locate two nails at the end of each terminal strip.
- Woven valley shingles can also be laid over the lining by first covering each roof area to a point approximately 3 feet from the center of the valley and weaving the valley shingles in place later.

Closed Cut Valleys

- This design can only be used with strip-type shingles and roll roofing materials.
- To install flashings at closed-cut valley locations, center a 36 inch wide strip of No. 15 asphalt-saturated, non-perforated felt in the valley over the existing No. 15 asphalt-saturated felt underlayment.
- Use only enough nails to hold the sheet smoothly in place.
- Lay the first course of shingles along the eaves of one roof area up to and over the valley.
- Extend it along the adjoining roof area for at least 12 inches.
- Continue to apply the second and successive courses over the valley in the same manner as the first course.
- Press the shingles into the valley and nail in place.
- Locate no nail closer than 6 inches from the valley's centerline.
- Locate two nails at the end of each terminal sheet (Figure 7.10).
- Apply the first course of shingles along the eaves of the intersecting roof area and extend it over the previously applied shingles.

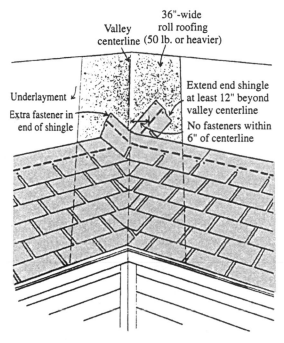

FIGURE 7.9 Weaving valley shingles.

- Trim a minimum of 2 inches back from the centerline of the valley.
- Clip the upper corner of each end shingle to prevent water from penetrating under the courses.
- Embed them in a 3 inch wide strip of plastic asphalt cement.

Flashing Other Roof Structures

- All structures that protrude above the roofline, including chimneys, vertical building walls, and dormers, must be carefully flashed at their bases.
- Use individual flashings, one for each course of shingles, along the sides that are perpendicular to the shingle courses.
- Bend each flashing at a right angle so that part of the flashing lies flat on the roof on top of the shingle, while the other half rests against the side of the chimney or dormer.

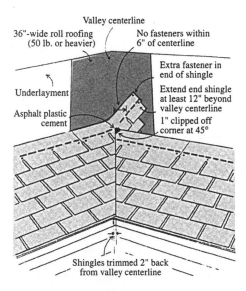

FIGURE 7.10 Closed-cut valley; two nails.

- Embed the upper edge in a mortar joint or cover with the dormer finish. Flash and counterflash the upper and lower sides, with one flashing running under the shingles and the other lying on top of them.

Flashing Against Side Vertical Walls

- Roof planes that butt against vertical walls at the end of shingle courses should always be protected by metal flashings that are protected by paint or other finishes if they are not made of a weather resistant metal like copper.

- These joints from the roof to vertical walls are among the most common sources of water intrusion into buildings. They must be designed correctly, installed correctly in unison with the other trades, and they must be maintained with diligence. Make them a part of your maintenance contracts.

- The method of protecting these intersections is called step flashing.

- The metal flashing shingles are rectangular in shape, from 5 to 6 inches long, and wider than the exposed surface of the roofing.

- Step flashing must be positioned over the shingle so that the next course of shingles covers it completely. The siding serves as counterflashing over step flashing.

- Study Figure 7.11 carefully, discuss these joints with the architect, and pay attention to the manufacturer's suggestions. If there are any questions, call for a meeting with the owner, the project manager, the related trades, and the architect.

- Bend the flashing shingle so that it extends 2 inches out over the roof deck.

- Extend the remainder of the shingle up the wall surface.

- Place each flashing shingle just above the exposed edge of the shingle that overlaps it, and secure it to the wall sheathing with one nail in the top corner, as shown in Figure 7.11.

- Because the metal is 7 inches wide and the roof shingles are laid 5 inches to the weather, each flashing shingle overlaps the next by 2 inches.

- Bring the finish siding down over the flashing to serve as cap flashing, but hold it far enough away from the shingles so that the ends of the boards can be readily painted to keep the roofing materials from becoming damp and deteriorating.

- If other trades are involved, discuss how these joints will be executed with the other mechanics.

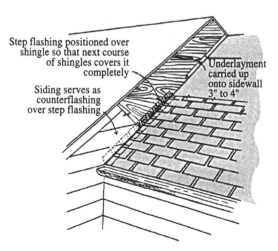

FIGURE 7.11 Flashing against side vertical walls.

Flashing Against Vertical Front Walls

- Apply the shingles up the roof until a course must be trimmed to fit at the base of the vertical wall.
- Plan ahead and adjust the exposure slightly in the preceding two courses so that the last course is at least 8 inches wide.
- Apply a continuous piece of metal flashing over the last course of shingles by embedding it in asphalt plastic cement and nailing it to the roof.
- Bend the metal flashing strip so that it extends at least 5 inches up the vertical wall and at least 4 inches onto the last shingle course.
- Do not nail the strip to the wall. Apply an additional row of shingles over the metal flashing strip, and trim it to the width of the strip.
- Then bring the siding down over the vertical flashing to serve as cap flashing.
- Place wooden siding far enough away from the roof shingles to allow for painting.
- Do not nail siding into vertical flashings. This is a prime source of the loss of roof integrity and the beginning of water intrusion.
- If the vertical front wall meets a sidewall, as in dormer construction, cut the flashing so that it extends at least 7 inches around the corner.
- Then continue up the sidewall with the step flashing.

Flashing Against Chimneys

- Chimneys are a very critical part of any roof. They must be treated carefully because they are a separate structure from the rest of the building.
- In order to accommodate stresses and distortions due to the uneven settling of roofing materials, the chimney usually sits on a separate foundation from the building and is normally subject to some differential settling.
- Therefore, flashing at the point where the chimney projects through the roof requires a type of construction that allows for movement without damaging the water seal.
- To satisfy this requirement, use base flashings that are secured to the roof deck and counterflashings that are secured to the masonry

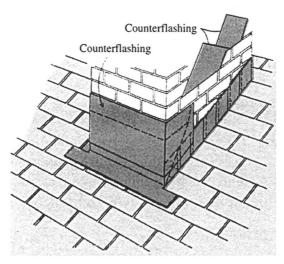

FIGURE 7.12 Flashing against chimney.

(Figure 7.12). With this technique, movement between the two structures does not destroy the integrity of the roof.

- Before any flashings are installed, apply shingles over the roofing felt up to the front face of the chimney and construct a cricket or saddle between the back face of the chimney and the roof deck.
- Design the cricket to prevent any accumulation of snow and ice and to deflect water around the chimney (Figure 7.13).
- Begin the flashing construction by installing at least 26 gauge galvanized metal, or an equivalent stain and corrosion resistant material, between the chimney and the roof deck on all sides.
- Apply the base flashing to the front first.
- Bend the base flashing so that the lower section extends at least 4 inches over the shingles and the upper section extends at least 12 inches up the face of the chimney.
- Set both sections of the base flashing in plastic asphalt cement.
- Use metal step flashing for the sides of the chimney.
- Position the pieces in the same manner as flashing for a vertical sidewall.
- Secure each piece to the masonry with plastic asphalt cement.

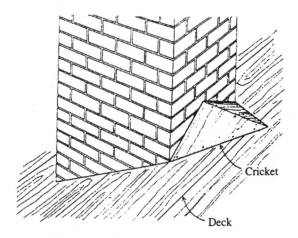

FIGURE 7.13 Location and configuration of chimney cricket.

- Use nails to secure the flashing to the roof deck.
- Embed the overlapping shingles in plastic asphalt cement.
- Place the rear base flashing over the cricket and the back of the chimney.
- Place counter-flashings over all apron, cricket, and step flashings to keep water out of the joint.
- Begin by carefully setting the metal counter-flashing into the brickwork as shown in Figure 7.14.
- This is done by raking out the mortar joint to a depth of 1½ inches and inserting the bent edge of the flashing into the cleared joint.

fastfacts

The metal base flashing should cover the cricket and extend onto the roof deck at least 6 inches. It should also extend 6 inches up the brickwork. Bring the asphalt shingles up to or over the cricket and cement in place.

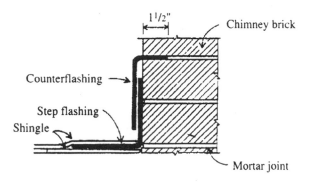

FIGURE 7.14 Metal counter-flashing into brickwork.

- Once it is in place and has a slight amount of spring tension, the flashing cannot be dislodged easily.
- Refill the grout joint with Portland cement mortar.
- Finally, bend the counter flashing down to cover the step flashing and to lie snugly against the masonry.
- Pay close attention to vent pipes, ventilators, and any other plumbing and mechanical openings through your roofs. They can easily become a source of water intrusion.
- All of these protrusions require special flashing methods that must be executed with care.
- If there are any questions, discuss them with the architect. Pay close attention to manufacturer's instructions.
- Do not proceed with your own work until all of the other trades have completed their work correctly.

Flashing Soil Stacks and Vents

- Apply shingles up to the vent pipe.
- Cut a hole in a shingle to go over the pipe and set the shingle in asphalt plastic roofing cement (Figure 7.15a).
- Examine a preformed, snug fitting flashing flange designed to go over the shingle and vent pipe.
- Prepare to seat the flange in asphalt plastic cement.

- When the flange is placed over the pipe, it must seat flatly on the roof (Figure 7.15b).
- After the flashing is in place, resume shingle application.
- Cut shingles in successive courses to fit around the pipe and embed them in asphalt plastic roofing cement where they overlay the flange.
- Avoid excessive use of cement as it can cause blistering.
- Do not drive fasteners close to the pipe.
- The completed installation should appear as shown in Figure 7.15c, with the lower part of the flange overlapping the lower shingles and the side and upper shingles overlapping the flange.
- Follow similar procedures when a ventilator or exhaust stack is located at the ridge.
- The typical difference is that the mechanic brings the shingles up to the pipe from both sides.
- Then, the flange is bent over the ridge so that the base works with both roof planes and overlaps the roof shingles at all points.
- Position the ridge shingles to cover the flange.
- Embed the ridge shingles in asphalt plastic cement where they overlap the flange.

Eave Flashing or Ice Dams

- Whenever there is the possibility of icing along the eaves, cement all laps in the underlayment courses from the eaves to a point at least 24 inches beyond the interior wall line of the building.
- The cemented double-ply underlayment serves as the eave flashing (Figure 7.16).
- To construct the eave flashing, cover the entire surface of the starter strip with a continuous layer of asphalt plastic cement applied at the rate of 2 gallons per 100 square feet.
- As an alternative, lap cement can be used at the rate of 1 gallon per 100 square feet.
- Place the first course over the starter course, and press it firmly into the cement.
- After the first course is in place, coat the upper 19 inches with cement.

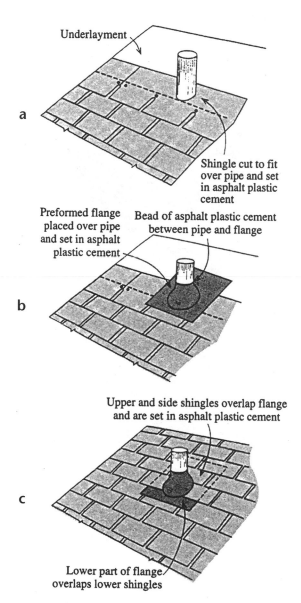

FIGURE 7.15 (a) Flashing soil stacks and vents. (b) flange over pipe. (c) complete installation.

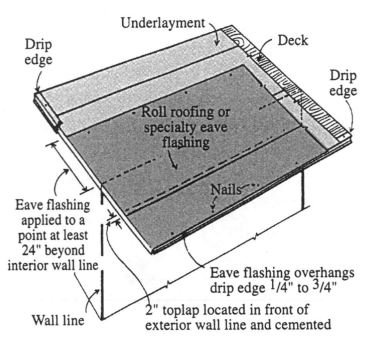

FIGURE 7.16 Eave flashing.

- Position the second course and press it into the cement.
- Repeat the procedure for each course that lies within the eave flashing distance.
- After completing the eave flashing, secure each successive course by using only enough fasteners to hold it in place until the shingles are applied.
- It is important to apply the cement uniformly so that the overlapping felt floats completely on the cement without touching the felt in the underlying course.
- Avoid excessive use of cement; it can cause blistering.

Drip Edges

- Drip edges are simple but they are very important, so apply them with care.

- The drip edge provides efficient water shedding at the eaves and rakes and helps to keep the underlying wood or plywood deck from rotting.
- Use 28 gauge galvanized metal (minimum thickness) or an equivalent corrosion and stain-resistant product to make-up the drip edges along eaves and rakes and at the rim of the decks.
- Provide an underlayment between the metal edge and the roof deck along the rake and over the metal edge along the eave.
- Extend the drip edge back from the edge of the deck not more than 3 inches.
- Secure the product with the appropriate nails, according to manufacturer's specifications, information, and codes.
- Space nails 8 to 10 inches apart along the inner edge.
- In high-wind areas, space nails 4 inches on center.

Finishing Hips and Ridges

- For hips and ridges, use individual shingles cut down to 12 x 12 inches from 12 x 36 inch three-tab shingles or to a minimum of 9 x 12 inches on two-tab or no-cutout asphalt shingles.
- Taper the lap portion of each cap shingle slightly so that it is narrower than the exposed portion.
- Some shingle manufacturers supply ready-cut hip and ridge shingles and specify how they should be applied.

fastfacts

➤ *Remember that galvanized metals and most other flashing products do deteriorate with time.*

➤ *Valleys and gutters (which are often butted up at the drip edge), even if they are made of copper, can fill up with debris. The standing debris can undermine the best flashing jobs by rotting the adjoining materials.*

➤ *Maintenance is very important on all roofs. Sell this service to clients rather than battling lawsuits.*

- To apply the cut ridge shingles, bend each shingle along its centerline so that it extends an equal distance on each side of the hip or ridge.

- Chalk lines can assist in proper alignment. In cold weather, warm the shingle until it is pliable before bending.

- Apply the shingles with a 5 inch exposure, beginning at the bottom of the hip or from the end of the ridge opposite the direction of the prevailing winds.

- As illustrated in Figure 7.17, secure each shingle with one fastener on each side, placed 5½ inches back from the exposed end and 1 inch up from the edge.

- The fastener length for hip and ridge shingles should be carefully determined from the maufacturer's instructions or the architect. This is a critical area that can cause callbacks and lawsuits.

Applying Strip Shingles on Low Slopes

- Asphalt strip shingles can be used on slopes between 2 and 4 inches per foot if special procedures are followed.

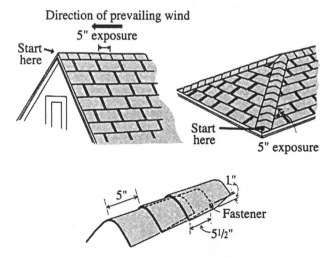

FIGURE 7.17   Secured shingle (special drawing of hip and ridge).

- Never use shingles with a rise lower than 2 inches in 12 inches.
- Water drains slowly from low slopes and creates a greater possibility of water backup and damage from ice dams.
- The special application method described below ensures that the low-slope roof remains weather-tight.
- Any of the shingle application methods discussed earlier in this chapter can be used on low slopes.

Underlayment

- On low-slope applications, cover the deck with two layers of asphalt saturated felt.
- Fasten a 19 inch wide strip of underlayment and place it along the eaves so that it overhangs the drip edge by ¼ to ⅜ inch.
- Place a full 36 inch wide sheet over the starter strip with the long edge placed along the eaves, completely overlapping the starter strip.
- Position succeeding courses of 36 inch wide sheets so that they overlap the preceding course by 19 inches.
- Secure each course using only enough fasteners to hold it in place until the shingles are applied.
- End laps should be 12 inches wide and located at least 6 feet from the end laps in the preceding course.

Eave Flashing

- Follow the instructions for applying eave flashing in the "Applying Flashing" section in this chapter.
- Special eaves flashing membranes also are available.
- If one of these specialty products is used, follow the manufacturer's application instructions with care.

Shingle Application

- For increased wind resistance, either use self-sealing shingles that incorporate a factory-applied adhesive or cement the tabs of free-tab shingles to the underlying course.

- To cement a free-tab shingle to the underlying course, you place two spots of asphalt plastic cement about the size of a quarter under each tab.
- Then press the tab into the adhesive.
- Be sure to cement all tabs throughout the roof.

REROOFING WITH STRIP SHINGLES

Many of the procedures used to apply shingles on new roofs also are followed in reroofing. Other procedures, specifically designed for reroofing applications, have been developed to meet the particular requirements of this type of work.

The differences depend primarily on the type of material that is on the existing roof, its condition, and whether or not the new roofing can be placed over it.

- Before reroofing, inspect the condition of the old roof and its supports.
- If the roof deck or its supports are found to be warped, rotted, or otherwise unsound, remove the old roofing and repair or replace the roof structure.
- If necessary, repoint the chimney and realign, clean, rebuild, or replace damaged gutters.
- Install flashing in valleys wherever needed and check that ventilation under the roof is adequate.
- Apply metal drip edges along the eaves and rakes.

In general, roofs covered with wooden shingles, asphalt shingles, asphalt roll roofing, slate, or built-up asphalt (BUR) roofing can be left in place. Depending on local codes, a maximum of three roofs, an original and two reroofs, can be installed before a tear-off becomes mandatory.

Tile or cedar shake roofs must be removed because of their irregular surface and the difficulty in fastening through them. The main consideration, however, is the condition of the framing beneath the existing roof, because new material installed on top of the old adds weight to the structure. The roof framing must be strong enough to support these additional loads plus the weight of the roofers and their equipment.

- First, inspect the deck to determine if it is sufficient for the new roof.

- Then, make repairs to the existing roof framing where needed to level it and to provide adequate strength.

- Remove all rotted or warped sheathing and delaminated plywood.

- Replace with new sheathing.

- Fill in all the spaces between boards with securely nailed wooden strips of the same thickness as the old deck or remove existing sheathing and re-sheath the deck.

- Pull out all protruding nails and re-nail the sheathing firmly at the new nail locations.

- If the deck consists of spaced sheathing, fill in all spaces with new boards of the same thickness as the old deck or cover the entire area with decking of the type and thickness required by local codes or the shingle manufacturer.

- Cover all large cracks, slivers, knotholes, loose knots, pitchy knots, and excessively resinous areas with sheet metal securely nailed to the sheathing.

- Sweep the deck clean again.

- Then follow the underlayment application directions described earlier in this chapter for new construction.

REROOFING OPERATIONS

- Preparatory procedures depend on the type of existing roofing.

- Five situations are generally encountered: asphalt shingles over old asphalt shingles, over wooden shingles, over roll roofing, over slate, and over BUR.

- Some shingles have a greater tendency to reveal the unevenness of the surface over which they are applied.

- This phenomenon is known as telegraphing, and can influence the choice of roofing material.

- If telegraphing might be a problem, consult individual manufacturers for advice before proceeding with the installation.

- Take measurements at both rakes.

- It is not unusual to find a difference of several inches between rakes on the same roof.

- Knowing this before you apply new shingles enables you to compensate for the difference in small increments over a series of courses.
- If the asphalt shingles with a 5 inch exposure are to be recovered with metric shingles, either a bridging or a nesting procedure is acceptable.
- Either procedure provides good results when properly applied.
- The bridging method, shown in Figure 7.18a, requires fewer shingles than the nesting procedure, illustrated in Figure 7.18b.
- Telegraphing of the underlaying material can result when the bridging method is used.
- Nesting, which minimizes an uneven appearance, is recommended when applying metric size shingles over an existing roof (Figure 7.19).
- For specific recommendations about bridging or nesting, contact the shingle manufacturer.

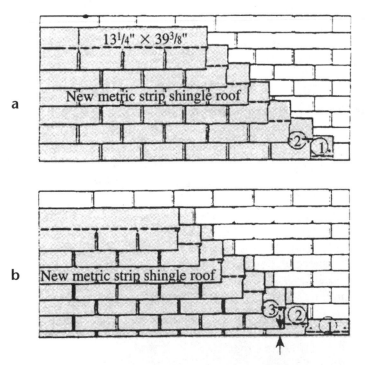

FIGURE 7.18 (a) Bridging method. (b) Nesting procedure.

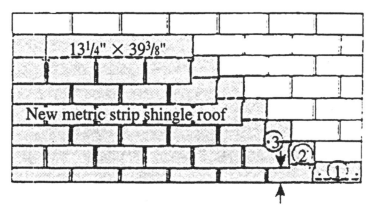

FIGURE 7.19 Nesting metric size shingles.

Reroofing Over Old Asphalt Shingles

- If the old asphalt shingles are to remain in place, remove any curled or lifted shingles and all loose and protruding nails.

- Remove all badly worn edging strips and replace with the new ones.

- Just before applying new roofing materials, sweep the old surface clear of all loose debris.

- If the old asphalt shingles are square-butt, and they are to be covered with new self-sealing, square-butt strip shingles, follow the application procedures in this section for applying shingles over an old roof.

- If the old asphalt shingles are locked down, stapled down, or badly curled strip shingles, the size, shape, and unevenness of the shingles can produce an uneven roof surface.

- New shingles tend to conform to the uneven surface.

- If a smoother surface is desired, remove the old shingles and prepare the deck using double underlayment.

Reroofing Over Old Wooden Shingles

- If an inspection of the roof indicates that the old wooden shingles can remain, carefully prepare the surface of the roof to receive the new roofing.

- Remove all loose or protruding nails, and renail the shingles in a new location.
- Nail down all loose shingles.
- Split all badly curled or warped old shingles and nail down the segments.
- Replace missing shingles with new ones.
- When the roof is subject to unusually high winds, cut back the shingles at the eaves and rakes far enough to allow the application at these points of 1 inch thick wooden strips that are 4 to 6 inches wide.
- Nail the strips firmly in place and allow the outside edges to project beyond the edge of the deck the same distance as the wood shingles.
- To provide a smooth deck, apply a backer board over the wooden shingles or use beveled wooden feathering strips along the butts of each course of old shingles.

Reroofing Over Old Roll Roofing

- When asphalt shingles are laid over old roll roofing, prepare the deck as follows.
- Slit all buckles and nail segments down smoothly.
- Remove all loose and protruding nails.
- If some of the old roofing has been torn away and sections of the deck are exposed, examine these areas to note any loose or pitchy knots and excessively resinous areas.
- Sweep the deck clean before applying new shingles.

Reroofing Over Old BUR

- BUR on a slope between 2 and 4 inches in 12 inches can be reroofed with asphalt shingles provided no insulation exists between the deck and the felts.
- Remove any old slag, gravel, or other material and leave the surface of the underlying felts smooth and clean.
- If a smooth, clean surface cannot be obtained, remove all of the old roofing and follow the procedures for reroofing when existing roofing is removed.

- Apply the asphalt shingles directly over the felts in the same manner as prescribed for new construction on low slopes.
- If rigid insulation exists under the BUR, remove the BUR and the insulation to expose the underlying deck.
- Then follow the low-slope application procedures outlined earlier in this chapter.

Reroofing Over Old Slate Roofs

- It is generally recommended that old slate roofs be removed before new asphalt shingles are applied, although it is possible to apply asphalt shingles successfully over this type of substrate by following the general recommendations described below and those suggested by the individual shingle manufacturer.
- The owner's architect must be consulted.
- The old roof slates must not be thicker than 3 inches.
- The roof deck must be solidly sheathed and in good condition.
- The slates must not shatter when nails are driven into them, regardless of placement.
- After years of exposure, most slates become quite soft, and needlepoint nails can be used with very little breakage or spalling.
- Begin by nailing down or removing any loose or projecting slates to provide a smooth surface for the new roofing material.
- Nails must be long enough to pass through the slates and penetrate the roof deck.
- Failure to follow these general application recommendations, especially those of the shingle manufacturer, can result in a poor roofing installation.
- Refer to local building codes for applicability and requirements regarding asphalt shingle applications over old slate roofs.

LAYING ASPHALT SHINGLES

Shingle application depends on whether the existing roofing has been removed or left in place, and on the type of roofing material left in place. If the old roofing has been removed or if reroofing is over old roll roofing or BUR, the shingle-application procedures are the same as those for new construction.

Wooden shingle roofs that are modified with feathering strips also are shingled in the same manner as new construction.

- The situation that differs from new construction is the application of new strip shingles over existing strip shingles.

- The nesting procedure, described earlier, minimizes any unevenness that might result from the shingles bridging over the butts of the old shingles.

- It also ensures that the new horizontal fastening pattern is 2 inches below the old one.

- The nesting procedure is based on the assumptions that the new roofing has a 5 inch exposure and the existing roofing was installed with a 5 inch exposure and properly aligned.

- If new eaves flashing has been added, snap chalklines on it to guide the installation of the new shingles until the courses butt against the existing courses.

- For the starter course, remove the tabs plus 2 inches or more from the top of the starter strip shingles so that the remaining portion is equal in width to the exposure of the old shingles, which is normally 5 inches.

- Apply the starter strip so that it is even with the existing roof at the eave.

- If self-sealing shingles are used for the strip, locate the factory-applied adhesive along the eaves.

- Be sure the existing shingles overhang the eaves far enough to carry water into the gutter. If they do not, cut the starter strip to a width that will do so.

- Do not overlap the existing course above.

- Remove 3 inches from the rake end of the first starter strip shingle to ensure that joints between adjacent starter strip shingles are covered when the first course is applied.

- For the first course, cut 2 inches or more from the butts of the shingles to be applied so that the shingles fit between the butts of the existing third course and the eave edge of the new starter strip.

- Start at the rake with a full-length shingle.

- Use four fasteners per shingle, locating them in the same positions as in new construction.

- Do not fasten into or above the factory-applied adhesive.

- Use full-width shingles for the second and all successive courses.

- Remove 6 inches from the rake end of the first shingle in each progressive course, until you reach the sixth course.
- Repeat the cycle by starting the seventh course with a full-length shingle.
- Place the top edge of the new shingles against the butt edge of the old shingles in the course above.
- Using the full width on the second course reduces the exposure of the first course to 3 inches, but gutters usually conceal this area and the appearance should not be objectionable.
- For the remaining courses, the 5 inch exposure is automatic and coincides with that of the existing shingles.
- As in new construction, apply the shingles across the diagonal as you proceed up the roof.

FLASHING AND REROOFING

Once the old roofing has been removed, flashing details generally follow those for new construction. If the existing flashings are still serviceable, however, they can be left in place and reused. If the old roofing is left in place, some flashing application procedures can differ from those used in new construction.

Vertical Walls

- If existing flashings are in good condition, reroof as follows.
- At a sidewall, trim ends of new shingles to within ¼ inch of the wall.
- Embed 3 inches of each shingle adjacent to the wall in plastic asphalt roofing cement.
- Finally, with a caulking gun, apply a thin line of roofing cement between the cut edge of the shingle and the wall.
- At a front wall, remove nailed tabs adjacent to the wall.
- Embed them in roofing cement and nail on the butt edge opposite the wall.
- Be careful not to nail within 3 inches of the vertical wall.
- Finish with a thin line of roofing cement between the cut edge of the shingle and the wall.

Valley Flashing

- If the existing roof has an open valley, build up the exposed area of the valley with mineral-surfaced roll roofing to a level flush with the existing roofing.
- Then install new open-valley flashing, in the same manner as for new construction, and overlap the existing shingles.
- The preferred treatment, however, is to construct a woven or closed-cut valley with the new shingles crossing over the valley filler strip.

Vent Pipes

- Carefully examine old metal flashing around all roof penetrations.
- If it has deteriorated, remove it and apply new flashing.
- If the metal flashing is in good condition, proceed as follows.
- Lift the lower part of the flange and apply the shingles underneath it up to the pipe.
- Replace the flange and embed it in asphalt roof cement.
- Protect the junction of the metal sleeve and the flange with an application of roof cement.
- Apply shingles around the pipe and up the roof.

Dormers

- A gable-type dormer is reroofed by using vertical sidewall and front wall methods, where applicable, and the closed-cut valley technique at the valleys.
- The roof of a shed dormer is shingled as a continuation of the main roof slope.

Chimneys

- Reuse serviceable metal chimney flashing, but don't take chances, if the old flashing has deteriorated—remove it.
- To do this, apply a strip of roll roofing approximately 8 inches wide on the reconditioned roof surface at the front and sides of the chimney.

- Lay the strip so that it abuts the chimney on all sides and secure it to the old roof with a row of nails along each edge.
- At the junction where it meets the chimney, apply a heavy coating of plastic asphalt cement to each course of shingles.
- Hold the lower edge slightly back from the exposed edge of the covering shingle, bend it up against the masonry, and secure it with suitable asphalt cement.
- Drive nails through the lower edge of the flashing into the roof deck.
- Cover these nails with the plastic cement used to secure the end shingle to the horizontal portion of the flashing and also the shingle itself.
- Repeat the operation for each course.
- The flashing units are wide enough to lap each other at least 3 inches, the upper one overlaying the lower one each time.
- Clean the masonry with a wire brush for a distance of 6 or 8 inches above the deck, and apply a suitable asphalt primer to the masonry surface.
- After the primer is applied, trowel plastic asphalt cement over the shingles for approximately 2 inches and up the chimney against the masonry surface for 4 to 6 inches.
- Press a strip of mineral-surfaced roll roofing wide enough to cover the cement into the cement.
- Return the sidepieces around and over the ends of the front piece and around the back of the chimney for a distance of about 6 inches.
- If the original construction did not provide a cricket behind the chimney, the base flashing at this point should consist of a 36 inch wide strip of mineral-surfaced roll roofing, applied to lie 24 inches up the roof, over the old roofing, and 12 inches up the rear face of the chimney.
- Embed the strip in plastic asphalt cement applied over the old roofing and against the primed masonry surface.
- Trowel the cement into all irregularities between the roof deck and the masonry.
- Secure the upper edge by nailing into a mortar joint.

APPLYING INDIVIDUAL SHINGLES

- Individual shingles are manufactured in three basic types: hexagonal, giant, and interlocking.
- Selection depends on many factors, including slope, wind resistance, coverage, aesthetics, and economics.
- Installation details vary according to the type of shingle used. The preparatory procedures are the same as those for applying strip shingles.

LAYING HEXAGONAL SHINGLES

- Two types of hexagonal shingles are available: those that lock together with a clip and those that have a built-in locking tab.
- Both the clip and lock-down shingles are relatively lightweight and intended primarily for placement over old roofing.
- They also can be used at times for new construction.
- For either application, the slope should be 4 inches per 12 inches or greater.
- Consult the roofing manufacturer for specific application instructions.

LAYING GIANT SHINGLES

- This type of shingle can be used for new construction or reroofing depending on the method of application.
- The Dutch lap method is intended primarily for reroofing over old roofing when a smooth surface and adequate anchorage is provided for nailing.
- It also can be used to cover new decks when single coverage provides the intended protection.
- For either application, the slope should be 4 inches per 12 inches or greater.
- Consult the roofing manufacturer for specific application instructions.
- The American method can be used for new construction or reroofing.

- In either case, the slope should be 4 inches in 12 inches or greater.
- Consult the roofing manufacturer for specific application instructions.

LAYING INTERLOCKING SHINGLES

- Interlocking shingles are manufactured with an integral locking device that provides immediate wind resistance.
- These shingles can be used for reroofing over existing roofing on slopes recommended by the shingle manufacturer.
- They can also be used for new construction depending on whether single or double coverage is required.
- In general, single-coverage interlocking shingles are not recommended for new construction.
- Check local building codes before installing interlocking shingles on new roofs.
- The location of the shingle fasteners is essential to the proper performance of the mechanical interlock.
- For best results, follow the shingle manufacturer's specifications concerning fastener placement.
- Follow the manufacturer's directions concerning application of the starter, first, and succeeding courses.
- Although interlocking shingles are self-aligning, they are flexible enough to allow limited adjustment.
- It is especially important to snap horizontal and vertical chalklines to keep the work in alignment.
- The integral locking tabs are manufactured within close tolerances to ensure a definite space relationship between adjacent shingles.
- Be sure to engage the locking devices carefully and correctly.
- Figure 7.20 illustrates two common locking devices used in interlocking shingles.
- During installation, locking tabs on shingles along the rakes and eaves might have to be removed in part or entirely.
- To prevent wind damage, shingles that have their locking tabs removed should be either cemented down or fastened in place according to the individual manufacturer's recommendations.

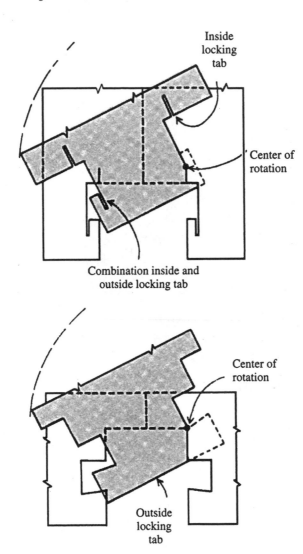

FIGURE 7.20   Two common locking devices.

APPLYING ROLL ROOFING

Asphalt roll roofing is used as both a primary roof covering and a flashing material. Roll roofing is applied using one of two different methods, depending on slope and other considerations.

- Asphalt roll roofing materials can be applied parallel to:
  - The rake using the exposed-nail method on roof decks having a slope of 4 inches in 12 inches or more.
  - The rake using the concealed-nail method on roof decks having a slope of 3 inches in 12 inches or more.
  - The eaves using the exposed-nail method on roof decks having a slope of 2 inches in 12 inches or more
  - The eaves using the concealed-nail method on roof decks having a slope of 1 inch in 12 inches or more.
- When maximum service life is an important consideration, use the concealed-nail method of application.
- As a primary roof covering, roll roofing is used on slopes as gradual as 2 inches in 12 inches.
- The roofing is applied either parallel to the eaves or parallel to the rakes.
- Nails should fit the application and have a length sufficient to penetrate ¾ inch into the deck or through the deck panel.
- Store the material on end in a warm place until ready for use, especially during the colder seasons of the year.
- It is not good practice to apply roll roofing when the temperature is below 45°F.
- If rolls do have to be handled below this temperature and have not been stored as suggested, warm them before unrolling them to avoid cracking the coating.
- Then cut the rolls into 12 to 18 foot lengths and spread them in a pile on a smooth surface until they flatten out.
- Before applying roll roofing, prepare the deck and install the necessary flashing in the manner described earlier for strip shingles.
- Use open valleys and follow the appropriate valley flashing procedures.
- Because all roll roofing is applied with a certain amount of top and side lapping, the proper sealing of the laps is crucial.

- Use only the lap cement or asphalt plastic cement recommended by the roofing manufacturer.
- Store the cement in a warm place until ready to use.
- The plastic cement is asphalt-based and contains solvents.
- Take proper safety precautions.
- Never heat asphalt cement directly over a flame.
- Do not attempt to thin the cement by diluting it with solvent.
- Always apply cement in a continuous but not excessive layer over the full width of the lap.
- Press the lower edge of the upper course firmly into the cement until a small bead appears along the edge of the sheet.
- Use a roller to apply pressure uniformly over the entire cemented area.
- Unless otherwise noted by the roofing manufacturer, apply asphalt plastic cement at the rate specified by the coating manufacturer.
- Excessive amounts of cement can cause blistering. During application allow sufficient time for volatiles to flash off.

APPLYING PARALLEL TO EAVES (CONCEALED-NAIL METHOD)

When using this method, place narrow edging strips along the eaves and rakes before applying the roofing material. Figure 7.21 illustrates the general installation procedure, including lapping, cementing, and nailing.

Edge Strips

- Place 9 inch wide strips of roll roofing along the eaves and rakes and position them to overhang the deck ¼ to ⅜ inch.
- Fasten the strips with rows of nails located 1 inch and 8 inches from the roof edge and spaced 4 inches on center in each row.

First Course

- Position a full-width strip of roll roofing so that its lower edge and ends are flush with the edge strips at the eaves and rakes.

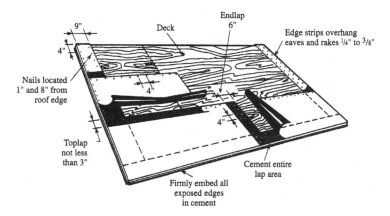

FIGURE 7.21 Roll roofing using a concealed-nail method.

- Fasten the upper edge with nails 4 inches on center and slightly staggered.
- Locate the nails so that the next course overlaps them a minimum of 1 inch.
- Lift the lower edge of the first course and cover the edge strips with cement according to the manufacturer's specifications.
- In cold weather, turn the course back carefully to avoid damaging the roofing material.
- Press the lower edge and rake ends of the first course firmly into the cement-covered edge strips.
- Work from one side of the sheet to the other to avoid wrinkling or bubbling.
- Endlaps should be 6 inches wide and cemented over the full lap area with the recommended cement.
- Nail the underlying sheet in rows 1 inch and 5 inches from the end of the sheet with the nails spaced 4 inches on center and slightly staggered.
- Endlaps in succeeding courses cannot line up with one another.

Second and Successive Courses

- Position the second course so that it overlaps the first course by at least 3 inches or as specified by the roofing manufacturer.

- Fasten the upper edge to the deck, cement the laps, and finish installing the sheet in the same manner as the first course.
- Follow the same procedure for each successive course.
- Do not apply nails within 18 inches of the rake until cement has been applied to the edge strip and the overlying strip has been pressed down.

Hips and Ridges

- Trim, butt, and nail the sheets as they meet at a hip or ridge.
- Next, cut 12 x 36 inch strips from the roll roofing and bend them lengthwise to lay 6 inches on each side of the joint.
- Do not bend the strips in cold weather without first warming them.
- Use these as shingles to cover the joint, each one overlapping the other by 6 inches, as shown in Figure 7.22.
- Start the hips from the direction opposite the prevailing winds.
- To guide the installation, snap a chalkline 5½ inches from and parallel to the joint on both sides.
- Apply asphalt plastic cement evenly over the entire area between the chalklines, from one side of the joint to the other.
- Fit the first folded strip over the joint and press it firmly into the cement.
- Drive two nails 5½ inches from the edge of the end that is to be lapped.
- Cover the 6 inch lap on this strip with lap cement.
- Then place the next strip over it.
- Nail and cement in the same manner as the first strip.
- Continue the same procedure until the hip or ridge is finished.

APPLYING PARALLEL TO EAVES (EXPOSED-NAIL METHOD)

Figure 7.23 illustrates the general installation procedures, including lapping, cementing, and nails.

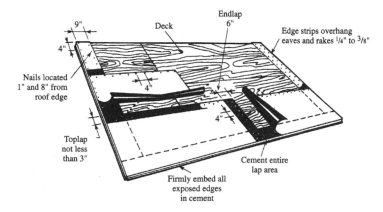

FIGURE 7.22 Concealed-nail method for hips and ridges.

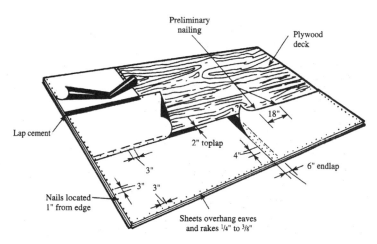

FIGURE 7.23 Application parallel to eaves using exposed-nail method.

First Course

- Position a full-width sheet so that its lower edge and ends overhang the eaves and rakes between ¼ and ⅜ inch.
- Nail along a line ½ to ¾ inch parallel to the top edge of the sheet. Space the nails 18 to 20 inches apart.
- This top nailing holds the sheet steady until the second course is placed over it and fastened.
- Nail the eaves and rakes on a line inset 1 inch and parallel to the edges of the roofing with the nails spaced 3 inches on center and staggered a bit along the eaves to avoid splitting the deck.
- If two or more sheets must be used to continue the course, lap them 6 inches.
- Apply lap cement to the underlying edge over the full lap width. Embed the overlapping sheet into it and fasten the overlap with two rows of nails 4 inches apart and 4 inches on center within each row.
- Stagger the rows so that the spacing is 2 inches between successive nails from row to row.

Second and Succeeding Courses

- Position the second course so that it overlaps the first course by 2 inches.
- Fasten the second course along the top edge following the same nailing directions as the first course.
- Lift the lower edge of the overlapping sheet and apply lap cement evenly over the upper 2 inches of the first course.
- Then embed the overlapping sheet into it.
- Fasten the lap with nails spaced 3 inches on center and staggered slightly.
- Place the nails not less than ¾ inch from the edge of the sheet.
- Nail the rake edges in the same manner as for the first course.
- Follow the same procedure for each successive course.
- Endlaps should be 6 inches wide and cemented and nailed in the same manner as for the first course.
- Stagger endlaps so that an endlap in one course is never positioned over the endlap in the preceding course.

Hips and Ridges

- Trim, butt, and nail the roofing as it meets at a hip or ridge.

- Snap a chalkline on each side of the hip or ridge, located 5½ inches from the joint and parallel to it.

- Starting at the chalklines and working toward the joint, spread a 2 inch wide band of asphalt lap cement on each side of the hip or ridge (Figure 7.24).

- Cut strips of roll roofing 12 inches wide and bend them lengthwise along the centerline so that they lay 6 inches on both sides of the hip or ridge.

- In cold weather, warm the material before bending it.

- Lay the bent strip over the joint and embed it in the cement.

- Fasten the strip to the deck with two rows of nails, one on each side of the hip or ridge.

- Locate the rows ¾ inch from the edges of the strip and space the nails 3 inches on center.

- Be sure the nails penetrate the 2 inch wide band of lap cement underneath in order to seal the nail hole with asphalt.

- Endlaps should be 6 inches and cemented the full lap distance. Avoid excessive use of cement, as it can cause blistering.

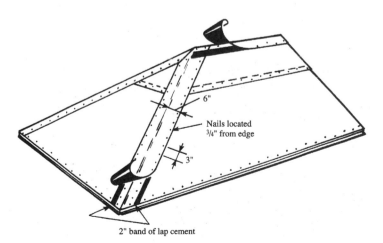

FIGURE 7.24 Hips and ridges using exposed-nail method.

APPLYING PARALLEL TO RAKE (CONCEALED-NAIL METHOD)

With this method, illustrated in Figure 7.25, sheets are applied vertically beginning at the eaves. Lay the sheets out and let them warm in the sun until they lie smoothly on a flat surface. If the sheets are nailed before they have time to relax, wrinkling can occur.

Edge Strips

- Place 9 inch wide strips of roll roofing along the eaves and rakes.
- Placement and nailing of these strips is identical to that for the concealed-nail method for applying parallel to the eaves.

First Course

- Position the first sheet so that it is flush with the edge strips at the rake and eaves.
- Fasten the upper edge with nails located 1 inch and 5 inches from the top edge of the sheet.
- Space the nails 4 inches on center.
- Carefully lift the first sheet back and cover both edge strips with an even layer of plastic cement.

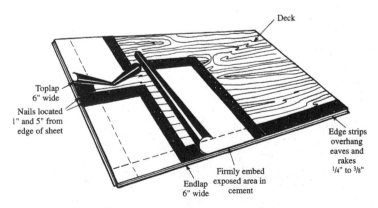

FIGURE 7.25 Applying parallel to rake using concealed-nail method.

- Excessive amounts of plastic cement can cause blistering.
- Press the sheet firmly into the cement.
- Work from the top of the sheet down to avoid wrinkling or bubbling.
- Endlaps should be a minimum of 6 inches wide.
- Fasten the length of the endlap with nails 4 inches on center and slightly staggered.
- Locate the nails so that the next sheet overlaps them a minimum of 1 inch.
- Cement endlaps over the full lap area with recommended lap cement.
- Top laps should be a minimum of 6 inches wide.
- Stagger the top laps of adjoining sheets to prevent a buildup where the sheets intersect.
- Cement endlaps over the full lap area with the recommended lap cement.

Second and Successive Courses

- Position the second course so that it overlaps the first course at least 6 inches.
- Fasten the upper edge to the deck, cement the laps, and finish installing the sheet in the same manner as the first course.
- Do not apply nails within 18 inches of the rake until cement has been applied to the edge strip and the overlying strip is pressed down.

Hips and Ridges

Finish the roof at these joints in the same manner as in the concealed nail method of application, parallel to the eaves.

APPLYING DOUBLE-COVERAGE ROLL ROOFING

- Double-coverage roll roofing is a 36 inch wide sheet, of which 17 inches are intended for exposure and 19 inches for a selvage edge.

- It provides double coverage for the roof and can be used on slopes as low as 1 inch per 12 inches.

- The 17 inch exposed portion is covered with granules, while the 19 inch selvage portion is finished in various manners depending on the manufacturer.

- Some vendors saturate the selvage portion with asphalt; some saturate it and coat it as well.

- Cement the selvage edge and all endlaps according to the manufacturer's recommendations.

- It is important to know the requirements of the particular product being used and to follow the roofing manufacturer's directions concerning the type and quantity of adhesive.

- Unless otherwise noted by the roofing material manufacturer, apply asphalt plastic cement at the rate of 2 gallons per 100 square feet of covered area or as recommended by the coating manufacturer.

- Make certain there is adequate roof drainage to eliminate the possibility of water standing in puddles.

- This is especially important on low slopes on which double-coverage roofing is commonly used.

- Choose the correct type and length of nail to fit the application. The fastener should be able to penetrate the deck ¾ inch or through the deck panel.

- Application of double-coverage roll roofing can be parallel to the eaves or to the rake.

- Although 19 inch selvage roll roofing is discussed here, any roll roofing can be applied in the same manner to obtain double coverage if the lapped portion of the sheet is 2 inches wider than the exposed portion.

- Before applying the roofing, prepare the deck and install flashings in the manner described earlier for strip shingles.

- Use open valleys and follow the appropriate valley flashing procedures.

APPLYING PARALLEL TO EAVES

- Remove the 17 inch granule-surfaced portion from a sheet of double coverage roll roofing.

- Place the remaining 19 inch selvage portion parallel to the eaves so that it overhangs the drip edge ¼ to ⅜ inch at both the eaves and rakes.

- Fasten it to the deck with two rows of nails, one on a line 4¾ inches from the top edge of the strip, the other on a line 1 inch above the lower edge.

- Space the nails 12 inches on center and stagger them in each row (Figure 7.26).

First Course

- Cover the entire starter strip with asphalt plastic cement.
- Avoid excessive use of cement, as it can cause blistering.
- Then position a full width sheet over it.
- Place the sheet so that the side and lower edge of the granule-surfaced portion are flush with the rake and eave edges of the starter strip.
- Fasten it to the deck with two rows of nails in the selvage portion.

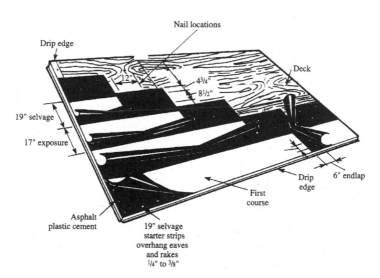

FIGURE 7.26 Applying parallel to eaves using starter strip.

- Locate the first row 4¾ inches below the upper edge and the second row 8½ inches below the first with the nails spaced 12 inches on center and staggered.

Successive Courses

- Position each successive course to overlap the full 19 inch selvage width of the course below and nail the selvage portion in the same manner as the first course.
- Turn the sheet back and apply cement to the full selvage portion of the underlying sheet.
- Turn the sheet back carefully to avoid damaging it.
- Spread the cement to within ¼ inch of the edge of the exposed portion.
- Press the overlying sheet firmly into the cement.
- Apply pressure over the entire lap using a broom or light roller to ensure complete adhesion between the sheets.
- It is important to apply the cement so that it flows to the edge of the overlying sheet under the application pressure.

Endlaps

- All endlaps should be 6 inches wide.
- Fasten the underlying granule surfaced portion of the lap to the deck with a row of nails 1 inch from the edge.
- Space the nails 4 inches on center.
- Then spread asphalt plastic cement evenly over the lap area.

fastfacts

➤ *Never cement roll roofing directly to the deck.*

➤ *This ensures that the sheets do not split due to deck movement.*

➤ *To make certain that roll roofing is not cemented to the deck when hot application is allowed, nail down a base sheet.*

- Embed the overlying sheet in the cement and secure the selvage portion of the sheet to the deck with nails on 4 inch centers in a line 1 inch from the edge of the lap.
- Stagger all endlaps so that those in successive courses do not line up with one another.

APPLYING PARALLEL TO RAKE

- With this method, the sheets are applied vertically from the ridge down.
- Begin by applying starter strips to both rakes using the procedures for horizontal application.
- Cover the starter strip with asphalt plastic and apply a full-width sheet over it as the first course.
- Position all endlaps so that the upper sheet overlaps the lower one and carries drainage over the joint rather than into it.
- The remainder of the application is the same as that for applying parallel to the eaves.
- Figure 7.27 shows the general arrangement for application parallel to the rake.

Hips and Ridges

- For both hip and ridge applications, trim, butt, and nail the roofing sheets as they meet at a hip or ridge.
- Snap chalklines 5½ inches from and parallel to the joint on each side to guide the installation.
- Next, cut 12 x 36 inch strips of roll roofing that includes the selvage portion.
- Bend the strips lengthwise to lie 6 inches on either side of the joint.
- In cold weather, be sure to warm the strips before bending.
- Start applying the strips at the lower end of the hip or at the end of the ridge opposite the direction of prevailing winds.
- Cut the selvage portion from one strip to use as a starter.
- Fasten this strip in place by driving nails 1 inch from each edge and 4 inches on center over the full length.
- Cover it completely with asphalt plastic cement.

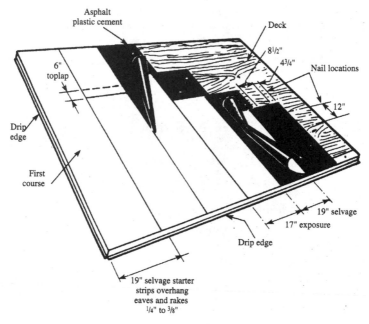

FIGURE 7.27 Application of double-coverage roll roofing parallel to rake.

- Fit the next folded strip over the starter and press it firmly into the cement, nailing it in the same manner as the starter, but only in the selvage portion.
- Continue the process until the hip or ridge is completed.
- Figure 7.28 illustrates the procedure.
- Double-coverage roll roofing is frequently used on sheds that contain no hips or ridges.
- To finish this type of roof, nail the selvage portion of the last course to the edge of the roof.
- Then trim the exposed, granule-surfaced portion that had been cut from the starter strip to fit over the final selvage portion and cement it in place.
- Finally, overlay the entire edge with metal flashing and cement it in place.

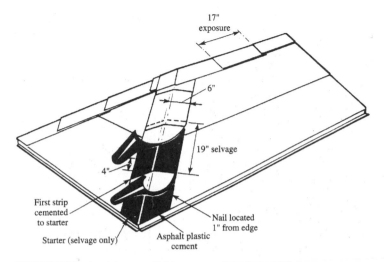

FIGURE 7.28 Applying parallel to eaves using starter strip.

INSPECTING THE JOB

- A quality installation is dependent on partial roof inspections during application and a final roof inspection upon completion.

- Ongoing review offers the roofer a number of opportunities to make certain that the materials are properly placed, properly fastened, and in good condition.

- Upon final inspection, give the roof a thorough cleaning.

- Remove any loose shingles, cuttings, nails, wood shavings, boards, or other debris that has been left lying on the roof.

- Remove any debris from valleys and gutters to ensure that they are unobstructed and can carry water away quickly and efficiently.

- Damaged shingles cause leaks and detract from the appearance of the finished job.

- Be sure to inspect the finished roof for any broken, torn, or damaged shingles.

- Remove damaged shingles by drawing the fasteners from the damaged shingle and the one immediately above it.

- Carefully fit a new shingle into place, fasten, and hand seal both shingles.

- Inspect all areas on which asphalt cement has been used to ensure that the materials have adhered well and that there are no bubbles in the roofing.
- If re-cementing is required to obtain a good bond, carefully lift the overlapping material to avoid damaging it, apply the cement, and work it thoroughly into the in-place material.
- Replace the overlying layer and press it firmly into the cement.
- Regardless of the type of asphalt roofing material used, check for proper fastening.
- Fasteners should not be exposed unless the installation procedures specifically call for exposure.
- No fastener should protrude from, or be driven into, the surface of the roofing material.
- Seat all protruding fasteners if they are placed improperly.
- A protruding fastener must be driven properly or replaced.
- Remove any bent or incorrectly placed fasteners as well as any that are driven too far into the surface of the asphalt roofing.
- Then pack the hole with asphalt plastic cement and drive a new fastener nearby.
- Before leaving the roof, examine the entire roof area for overall appearance and any defects that might have been overlooked at closer range.
- Thoroughly clean the site of roofing materials, cuttings, scraps, wood, and any other debris remaining from the roofing work.
- If possible, before leaving the job, have the owner make a final inspection of the grounds and obtain his or her approval of the work.

RESOLVING PROBLEM AREAS

Two problem areas that can result in callbacks are shading and algae discoloration. Neither of these problems is the result of poor workmanship.

SHADING

- As a completed asphalt shingle roof is viewed from different angles, certain areas can appear darker or lighter.

- This difference in appearance is called shading.
- Shading also depends on the position of the sun and the overall intensity of light.
- For example, slanting sunrays emphasize shading, while direct, overhead rays cause color shading to disappear.
- Shading is a visual phenomenon that in no way affects the performance of the shingles.
- It occurs primarily as a result of normal manufacturing operations that produce slight differences in surface texture that cannot be detected during the production process.
- These unavoidable variations in texture simply affect the way the surface reflects light.
- Shading is usually more noticeable on black or dark-colored shingles because they reflect only a small part of the light shining on them.
- This magnifies the slight differences in surface texture.
- White and light-colored shingles reflect a great amount of light, which diminishes observable shading differences.
- Blends made from a variety of colors actually tend to camouflage shading, with lighter colored blends reducing the effect of shading more than darker blends.
- Racking or straight-up application accentuates shading.
- For this reason, be sure shingles are applied across and diagonally up the roof.
- This blends the shingles from one bundle into the next and minimizes any shade variation from one bundle to the next.
- Shading can also occur when the backing material used to keep shingles from sticking together in the bundle rubs off onto the exposed portion of a shingle.
- Staining can also appear in shingles that have been stacked too high or for too long a period of time.
- This type of shading develops because of minor staining from the oils in the asphalt coating.
- In either of these cases, weathering usually eliminates the problem over a period of time.

ALGAE DISCOLORATION

- Roof discoloration caused by algae, commonly referred to as fungus growth, is a frequent problem throughout the country.
- It is often mistaken for soot, dirt, moss, or tree droppings.
- The algae that cause this discoloration do not feed on the roofing material and, therefore, do not affect the service life of the roofing.
- The natural pigments in the algae, however, can gradually turn a white or light roof dark brown or black over a period of years.
- Algae stains are difficult to remove from roofing surfaces, but the coloration can be lightened with a diluted cleaning solution.
- Contact the appropriate manufacturer for details.
- Gently sponge the solution onto the roofing material.
- Scrubbing loosens and removes granules, so apply the solution gently to avoid damaging other parts of the building and the surrounding landscape.
- If possible, work from a ladder or walk boards to avoid walking directly on the roof. Observe safety precautions whenever working on or near the roof.
- After sponging, rinse the solution off the roof with a hose.
- Observe strict safety precautions, as this process makes the roof slippery and potentially hazardous during treatment.
- The effectiveness of such cleaning is only temporary, and the discoloration can recur.
- Several types of algae-resistant roofing have been developed and are now commercially available.
- These asphalt roofing products are specifically designed to inhibit most algae growth for extended periods of time.
- When undertaking an asphalt roofing job in an area where algae discoloration is a problem, advise the property owner of available algae-resistant asphalt roofing products.

8

WOOD SHINGLES

Wood shingles and shakes are most often cut from red cedar, redwood, cypress, or pine logs. Western red cedar is by far the most common shingle lumber used in the United States.

The cellular composition of Western red cedar, with millions of tiny, air-filled cells per cubic inch, provides thermal insulation on both roof and wall applications as well as with shedding water.

Cedar also has outstanding rigidity in hurricane winds and demonstrates great resilience under the pounding of hail. These natural advantages are amplified by the structural strength derived from the overlapping method typically used for application.

Most cedar roofs are not chemically treated. Solutions are used, however, to get the best service in climatic areas that experience a combination of heat and humidity. Check local building codes for conformity with these recommendations.

To make shingles and shakes, selected logs are cut to specified lengths and the resulting blocks or bolts are trimmed to remove bark and sapwood. Then shingles are sawn from the bolts, while shakes can be either hand-split or sawn (Figure 8.1).

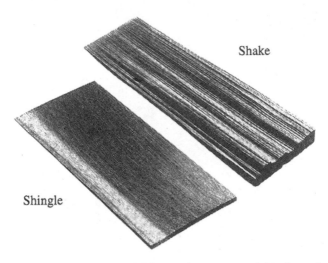

FIGURE 8.1   The physical difference between wood shingles and shakes.

GRADING SHINGLES AND SHAKES

Grading Shingles

Like lumber and plywood, wood shingles are graded and labeled to indicate the quality of the product. The Cedar Shake and Shingle Bureau grades red cedar shingles. The thickness at the butt end is used to grade shingles by thickness. For example, 5/2 means that 5 butts stacked on top of each other are 2 inches thick.

Number 1 Grade, Blue Label Shingles

- The premium grade of shingle is No. 1, or blue label.
- These shingles are cut tapered and smooth sawn on both faces for a neat, tailored appearance. They are cut to one of three lengths:
  - 16 inches
  - 18 inches
  - 24 inches
- No. 1 shingles are clear heartwood that have no defects and are 100 percent edge grain.

- On 4/12 and steeper pitched roofs, a No. 1 shingle is applied at:
  - 5 inch exposure for a 16 inch shingle.
  - 5½ inches for an 18 inch shingle.
  - 7½ inches for a 24 inch shingle.

Number 2 Grade, Red Label Shingles

- A No. 2, or red label, shingle is cut in the same lengths and thicknesses as the No. 1 shingle.
- The face must be:
  - 10 inches clear on the 16 inch shingles.
  - 11 inches clear on 18 inch shingles.
  - 16 inches on 24 inch shingles.
  - Limited sapwood and flat grain are allowed.
  - Limited knots and defects are allowed above the clear portion.
- On a 4/12 or steeper pitched roof, No. 2 shingles are applied at:
  - 4 inches for 16 inch shingles.
  - 4½ inches for 18 inch shingles.
  - 6½ inches for 24 inch shingles.
  - No. 2 shingles are most often used as a starter course and for reroofing.

Number 3 Grade, Black Label Shingles

- Utility grade for economy applications and secondary buildings.
- A No. 3 grade shingle is cut in the same lengths and thicknesses as the No. 1 and No. 2 shingles.
- The face must be 6 inches clear on the 16 and 18 inch shingles and 10 inches clear on 24 inch shingles.
- Sapwood and flat grain are allowed.
- Limited knots and defects are permitted above the clear portion.
- On a 4/12 or steeper pitched roof, No. 3 shingles are applied at:
  - 3½ inch exposure for 16 inch shingles.
  - 4 inch exposure for 18 inch shingles.
  - 5½ inch exposure for 24 inch shingles.

Number 4 Grade, Utility Shingles

Used for starter courses and undercoursing.

Grading Shakes

The four types of shakes are categorized by their texture and by how they are manufactured.

Tapersawn Shakes

Tapersawn shakes are available in three types:

- No. 1:
  - Both faces are sawn like that of a shingle, but it is cut thicker.
  - This shake presents the naturally tailored look of a wooden shingle roof, but the thicker butt creates the complimentary sharp shadowline of a shake.
  - Cut from clear heartwood.
  - 100 percent edge grain with no defects.
  - This shake is ⅝ inches thick at the butt.
  - Cut in two lengths: 18 and 24 inches.
  - On a 4/12 or steeper pitched roof, this product is exposed at 7½ inches for 16 inch shakes and 10 inches for 24 inch shakes.
- No. 1 grooved tapersawn:
  - This is a machine-grooved shake.
  - It resembles a hand-split or resawn shake.
  - The grooved tapersawn shake is applied at the same exposure as the No. 1 tapersawn, above.
  - As with the tapersawn shake, a ¾ inch thick butt can be custom cut for an even sharper shadow line.
  - The exposure application is the same as that for the ⅝ inch thick tapersawn shake.
- The No. 2 tapersawn:
  - Is cut in the same lengths and thicknesses as the No. 1 tapersawn shake.
  - The lower half of the shake is clear.
  - Flat and cross grains are allowed.

- The top half of the shake is allowed tight knots and other limited defects.

- The exposure for this lower grade of shake is dropped from 7½ inches to 5½ inches for the 18 inch shake and from 10 inches to 7½ inches for the 24 inch shake.

- This product is used for reroofing residences and for wall shake applications, starter coursing, sheds, etc.

The Hand-Split and Resawn Shake

The hand-split and resawn shake are available in two types:

- The heavy hand-split and resawn shake has a sawn back face.
  - The face is hand-split so that the natural grain of the wood gives it a highly textured look.
  - It is also heavier than any other shake, which gives the roof a rugged appearance.
  - This shake is cut from clear heartwood and 100 percent edge grain, with no defects.
  - It is 1½ inches thick at the butt.
  - It is cut in two lengths: 18 and 24 inches.
  - On a 4/12 or steeper pitched roof, it is applied at 7½ inches for 18 inch shakes and 10 inches for 24 inch shakes.
- The heavy hand-split and resawn shake:
  - Cut in the same manner as the medium hand-split but is not cut as thick.
  - It still gives a very textured appearance, as it is made from clear heartwood and 100 percent edge wood with no defects.
  - This shake has a ½ inch thick butt end.
  - It is cut in two lengths: 18 and 24 inches.
  - It is applied with the same exposure as the heavy shakes.
- The tapersplit shake:
  - It is produced by reversing the block, end for end, with each split.
  - The best part of the tree is required for this naturally split shake, which is largely produced by hand with a mallet and froe.
  - A tapersplit roof gives a less rustic, rippled shadowline.
  - It is cut from clear heartwood, 100 percent edge grain, and has no defects.

- The tapersplit shake is cut ½ inch thick and 24 inches in length.
- It is applied at 10 inch exposure on a 4/12 or steeper pitched roof.
- Straightsplit shakes:
  - They are also known as barn or ranch shakes.
  - They are commonly mistaken for a product exclusively used on barns.
  - They are produced mainly by machine, but also can be split with a mallet and froe like tapersplit shakes.
  - The difference between the straightsplit and the tapersplit is that the straightsplit is split from the same end of the block, which produces the same thickness throughout.
  - The straightsplit creates a rippled shadowline on the roof.
  - They are cut from clear heartwood and are 100 percent edge grain, with no defects.
  - This shake is ⅜ to ½ inch thick.
  - Straight splits are cut in two lengths: 18 and 24 inches.

Treated Shakes

- No. 1, or blue label, shingles and No. 1 split shakes are available pressure-impregnated with fire retardants that meet testing standards developed by Underwriters' Laboratories, Inc. (UL) and adopted by the National Fire Protection Association (NFPA).
- These shingles and shakes also are warranted for 30 years by the treater and/or chemical company when they are 0.40 chromium copper alummate (CCA) pressure treated.
- Treated products are ideal in areas of high humidity where premature decay can occur.
- Specify the Certi-Last treating label for this extra protection.

ESTIMATING QUANTITIES

- To calculate the material needed for reduced exposures, divide the square footage by the reduced coverage.
- Table 8.1 shows that a typical four-bundle square of 16 inch (No. 1 blue label or No. 2 red label) shingles, at 4 inch exposure, covers 80 square feet.

- For 3200 square feet of surface, as the formula indicates, 40 squares of shingles (3200 divided by 80 = 40) are needed.
- To estimate shake coverage, use the formulas and information in Table 8.2.

APPLYING GENERAL DESIGN DETAILS

Use wooden shingles when the roof's slope or pitch is sufficient to ensure good drainage. Climatic conditions, the skill of the installer, and the application techniques are modifying factors when evaluating a proposed installation.

- The minimum recommended pitch is 1/6 or 4 in 12 (4 inch vertical rise for each 12 inch horizontal run), although there have been satisfactory installations on lesser slopes.
- For shakes, the recommended minimum slope is 3 inches vertical rise for every vertical foot.
- Shingles and shakes can be applied over spaced sheathing.
- Spaced sheathing is usually 1 x 4 or 1 x 6 inch softwood boards.
- Solid sheathing is acceptable and might be required in seismic regions or under treated shakes and shingles.
- It is recommended for shake applications in areas where wind-driven snow is common.
- Solid sheathing is usually strand-board panels or plywood, which provides a smooth base for roofing.

TABLE 8.1 Shingle Coverage

	Weather exposure							
	$3\frac{1}{2}''$	$4''$	$5''$	$5\frac{1}{2}''$	$6''$	$6\frac{1}{2}''$	$7''$	$7\frac{1}{2}''$
Length × thickness	Approximate coverage (in sq. ft.) of 1 square (4 bundles) of shingles							
$16'' \times 5/2$	70	80	90	100	—	—	—	—	—
$18'' \times 5/2\frac{1}{4}$	—	$72\frac{1}{2}$	$81\frac{1}{2}$	$90\frac{1}{2}$	100	—	—	—	—
$24'' \times 4/2$	—	—	—	—	$73\frac{1}{2}$	80	$86\frac{1}{2}$	93	100

TABLE 8.2 Shake Coverage

Length × thickness, shake type	Weather exposure				
	5″	5$^{1}/_{2}$″	7$^{1}/_{2}$″	8$^{1}/_{2}$″	10″
	Approximate coverage (in sq. ft.) of 1 square of shakes[a]				
18″ × $^{1}/_{2}$″, hand-split and resawn mediums[b]	—	55[c]	75[d]	—	—
18″ × $^{3}/_{4}$″, hand-split and resawn heavies[b]	—	55[c]	75[d]	—	—
18″ × $^{5}/_{8}$″, tapersawn	—	55[c]	75[d]	—	—
24″ × $^{3}/_{8}$″, hand-split	50[e]	—	75[c]	—	—
24″ × $^{1}/_{2}$″, hand-split and resawn mediums	—	—	75[c]	85	100[d]
24″ × $^{3}/_{4}$″, hand-split and resawn heavies	—	—	75[c]	85	100[d]
24″ × $^{5}/_{8}$″, tapersawn	—	—	75[c]	85	100[d]
24″ × $^{1}/_{2}$″, tapersawn	—	—	75[c]	85	100[d]
18″ × $^{3}/_{8}$″, straightsplit	—	65[c]	90[d]	—	—
24″ × $^{3}/_{8}$″, straightsplit	—	—	75[c]	85	100[d]

15″ starter-finish course: Use as supplemental with shakes applied not to exceed 10″ of weather exposure.

[a]All coverage based on an average $^{1}/_{2}$″ spacing between shakes.

[b]5 bundles cover 100 sq. ft. roof area when used as starter-finish course at 10″ weather exposure; 7 bundles cover 100 sq. ft. roof area at 7$^{1}/_{2}$″ weather exposure. See *a*.

[c]Maximum recommended weather exposure for 3-ply roof construction.

[d]Maximum recommended weather exposure for 2-ply roof construction.

[e]Maximum recommended weather exposure.

SHINGLE SHEATHING

There are two acceptable application methods for spaced sheathing:

- Using 1 x 4 inch boards.
  - Spread 1 x 4 inch boards to coincide with the weather exposure (Table 8.3) of the shingles (Figure 8.2).
  - In other words, if the shingles are to be laid at 5½ inches to the weather, the sheathing boards also should be spaced at 5½ inches on center.
  - In this method of application, each shingle is nailed to the center of the sheathing board.
- Using 1 x 6 inch boards.
  - Two courses of shingles are nailed to each 1 x 6 inch board (Figure 8.3), up to and including 5½ inches of weather exposure.
  - With 7½ inches of weather exposure, the center of the sheathing board equals the distance of the weather exposure.
  - Although not commonly used, a breather-type underlayment, such as roofing felt, can be applied over either solid or spaced sheathing.

SHAKE SHEATHING

- In shake applications, spaced lumber is usually 1 x 6 inch boards spaced on centers equal to the weather exposure (Figure 8.4) at which the shakes are to be laid.

TABLE 8.3 Shingle Exposure

Pitch	No. 1, blue label			No. 2, red label			No. 3, black label		
	Length (in.)								
	16	18	24	16	18	24	16	18	24
3/12–4/12	3¾	4¼	5¾	3½	4	5½	3	3½	5
4/12 and steeper	5	5½	7½	4	4½	6½	3½	4	5½

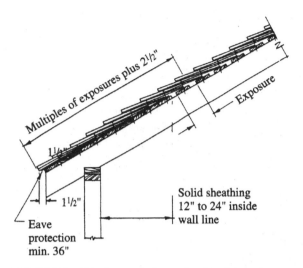

FIGURE 8.2 Shingles applied over 1 x 4 inch boards.

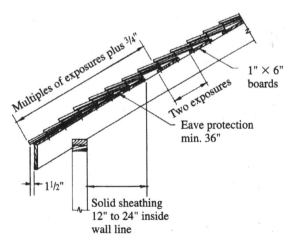

FIGURE 8.3 Shingles applied over 1 x 6 inch boards.

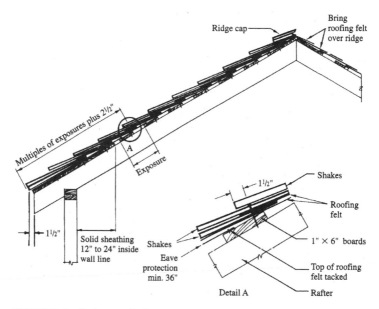

FIGURE 8.4 Shakes applied over 1 x 6 inch boards.

- The spacing should never be more than 7½ inches for 18 inch shakes and 10 inches for 24 inch shakes on the rooftop.

- When 1 x 4 inch sheathing is installed at 10 inches on center, install 1 x 4 inch boards between the sheathing boards.

- A solid deck is recommended in areas where wind-driven snow is a factor.

- Roofing felt interlay between the shake courses is required whether the sheathing is spaced or solid.

- The felt interlay acts as a baffle that prevents wind-driven snow or other foreign material from entering the attic cavity during extreme weather conditions.

- The interlays also increase the roof's insulating value.

- Take special care when installing the felt interlays over spaced sheathing to ensure the formation of an effective baffle.

- Apply the felt over the top portion of the shakes and extend it onto the spaced sheathing.

- Position the bottom edge of the felt at a distance above the butt end that is equal to twice the weather exposure.

fastfacts

To be an effective baffle, the top of the felt must rest on the sheathing. Place the shakes so that the nails are driven through the upper portion of the sheathing board and attach the tip of the roofing felt to the lower portion.

The following details must be observed on all applications (Figures 8.5 and 8.6.):

- Shingles must be doubled or tripled at all eaves.
- Project butts of first course shingles 1½ inches beyond the fascia.
- Spacing between adjacent shingles (joints) should be a minimum of ¼ inch and a maximum of ⅜ inch.
- Separate joints in any one course by not less than 1½ inches from joints in adjacent courses.
- In any three courses, no two joints should be in direct alignment.
- In lesser-grade shingles that contain both flat and vertical grain, do not align joints with the centerline of heart.
- Split flat grain shingles that are wider than 8 inches in two before nailing. Treat knots and similar defects as the edge of the shingle, and place the joint in the course above 1½ inches from the edge of the defect.

APPLYING SHAKES

Shakes, like shingles, are normally applied in straight, single courses. As with shingles, follow the application details (Figure 8.7).

- The starter course can be one or two layers of shingles or shakes overlaid with the desired shake.
- A 15 inch shake is made expressly for starter and finish courses.
- Project butts of first course shakes 1½ inches beyond the fascia.

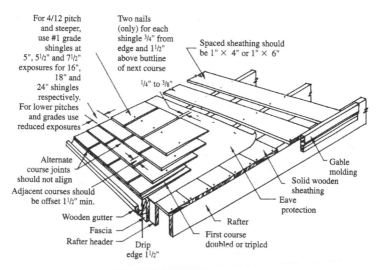

FIGURE 8.5 Shingle applications.

APPLYING SHINGLES	
	ITEM
	Shingles must be doubled or tripled at all eaves.
	Project butts of first course shingles 1-1/2 inches beyond the fascia.
	Spacing between adjacent shingles (joints) should be a minimum of 1/4 inch and a maximum of 3/8 inch.
	Separate joints in any one course by not less than 1-1/2 inches from joints in adjacent courses. In any three courses, no two joints should be in direct alignment.
	In lesser-grade shingles that contain both flat and vertical grain, do not align joints with the centerline of heart.
	Split flat grain shingles that are wider than 8 inches in two before nailing. Treat knots and similar defects as the edge of the shingle, and place the joint in the course above 1-1/2 inches from the edge of the defect.

FIGURE 8.6 Applying shingles.

- Lay an 18 inch wide strip of No. 30 roofing felt, or No. 15 felt (depending on specifications and code requirements) over the top portion of the shakes and extend it onto the sheathing.

- Position the bottom edge of the felt so that it is above the butt of the shake by a distance equal to twice the weather exposure.

- For example, apply the felt 20 inches above the butt for 24 inch shakes laid with 10 inches of exposure.

- The felt covers the top 4 inches of the shakes and extends 14 inches onto the sheathing.

- Note that the top edge of the felt must rest on the spaced sheathing.

- Spacing between adjacent shakes should be a minimum of ⅜ inch and a maximum of ⅝ inch.

- Offset joints between shakes 1½ inches over adjacent courses.

- Lay straightsplit shakes with the smoother end from which the shake was split, or froe end, toward the ridge.

- For low slope, apply a conventional hot-mop or suitable roll-type asphalt roof over the roof deck (see Chapter 5).

- With the final hot-mop application, embed 2 x 4 inch spacers of Western red cedar or preservative-treated lumber in the bituminous coating.

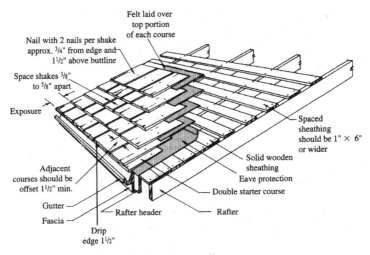

FIGURE 8.7 Shake application.

fastfacts

➤ *The minimum roof slope recommended for shakes is ⅓ and for shingles is ¼.*

➤ *It is possible, however, to apply shingles or shakes successfully to solid-sheathed roofs with lower slopes following a special application method (Figure 8.8).*

➤ *The special method provides a double roof on which the shingles or shakes are applied to a lattice-like framework that is embedded in a bituminous surface coating.*

- Install these spacers over the rafters at 24 inches on center and extend them from eave to ridge.

- Next, nail 1 x 4 inch or 1 x 6 inch nailing strips, spaced according to the weather exposure selected for the shingles or shakes, across the spacers to form a lattice-like nailing base.

- For example, if 24 inch shakes are to be installed at a weather exposure of 10 inches, the nailing strips also should be spaced at 10 inches on center.

- Finally, apply the shingles or shakes in the normal manner with a starter course at the eave and felt interlays between each course.

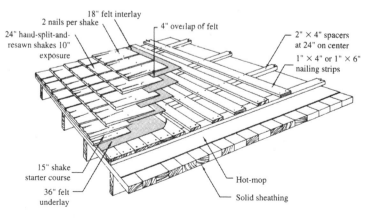

FIGURE 8.8 Application of shakes to low-slope roof.

Detailing Hips and Ridges

- Cap intersecting roof surfaces at the hip and ridge areas to ensure a weathertight joint.
- For the final course at the ridgeline, as well as for hips, select uniform shakes.
- Apply a strip of 15 pound roofing felt, at least 8 inches wide, over the crown of all hips and ridges.
- Always fabricate hips and ridges with shakes to create a harmonious appearance.
- Use prefabricated hip-and-ridge units, or fabricate the hips and ridges onsite.
- To fabricate hips onsite, sort out shakes that are approximately 6 inches wide.
- Tack two wooden straightedges onto the roof.
- Place one on each side, 4 to 5 inches from the centerline of the hip (Figure 8.9).
- Double the starting course of the shakes.
- Place the first shake on the hip with one edge resting against the guide strip.
- Cut the edge of the shake projecting over the center of the hip on a bevel.
- Then apply the shake on the opposite side and cut back the projecting edge to fit.
- Shakes in following courses are applied alternately in reverse order.
- Ridges are constructed in a similar manner.
- Weather exposure should be the same as that given the shakes on the roof.

fastfacts

It is important that shakes on hips and ridges be attached with nails of sufficient length to penetrate the underlying sheathing by at least ½ inch.

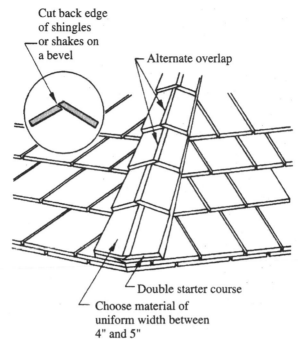

FIGURE 8.9 Hip and ridge application.

- When laying a cap along an unbroken ridge that terminates in a gable end, start laying the cap at each end so that it meets in the middle of the ridge.

- At that point, nail a small saddle of shake butts to splice the two lines.

- Always double the first course of capping at each end of the ridge.

- Conceal nails with the overlapping hip and ridge unit, as in the field of the roof.

Protecting Eaves

- Extend eave protection on all shingle and shake roofs from the edge of the roof to a line up the roof slope not less than 12 inches inside the inner face of the wall.

- This protection is not necessary if ice dams are not likely to form along the eaves and cause a backup of water.

- Eave protection is not required over unheated garages, carports, and porches, where the roof overhang exceeds 36 inches (measured along the roof slope) or where low-slope shingles are used.

Detailing Roof Junctures

- The correct construction of roof junctures ensures weather tightness.

- When metal flashing is employed, use nothing less than 26-gauge galvanized steel or an acceptable equivalent.

- Paint the metal on both sides with a good metal or bituminous paint.

- Paint flashing materials after bending; this helps maintain the integrity of the coating.

- For convex junctures (Figure 8.10), install metal flashings to cover the top 4 inches of the wall and the bottom 8 inches of the roof slope before the final course of shingles or shakes is nailed to the top of the wall.

- Apply a narrow, horizontal band of shingles or shakes, or a strip of wooden molding, after the final wall course is installed.

- Apply a double or triple starter course at the eave, with a 1½ inch overhang of the wall surface.

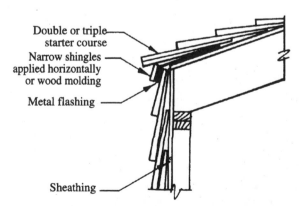

FIGURE 8.10 Convex roof juncture.

- Then complete the roof in the normal manner.
- Metal flashings for concave junctures (Figure 8.11) are similar to those for convex junctures.
- Install the metal flashings to cover the top of the roof slope and the bottom 4 inches of the wall before applying the final course of shingles or shakes.
- Install the final roof course so that the tips fit as snugly as possible against the wall at the juncture.
- Apply a double starter course at the base of the wall surface and then complete the remaining wall courses in the recommended manner.
- For apex junctures (Figure 8.12a), cover the top 8 inches of the roof and the top 4 inches of the wall with the metal flashing.
- Install the flashing before applying the final course of shingles or shakes to the wall.
- The recommended application sequence is to apply shingles or shakes first to the wall and then to the roof.
- Trim the overhanging roof material flush with the wall.
- Finally, apply specially prepared ridge units over the wall-roof juncture so that the roof piece overlaps the wall piece in each matching pair.
- When shingles or shakes are to be applied to a swept or bell eave (Figure 8.12b) with an excessive curvature, it might be necessary

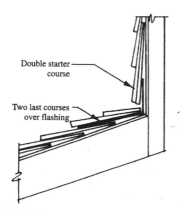
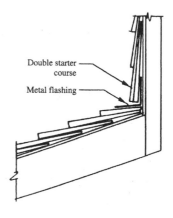

FIGURE 8.11 Concave roof juncture.

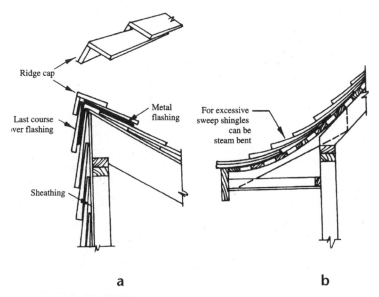

a   b

FIGURE 8.12 (a) Apex roof juncture. (b) Swept or bell eave.

to soak the shingles or shakes, usually overnight, or steam them prior to installation.

• Employ a double starter course in the usual manner.

• The slope of the roof and the type of shingle or shake selected determines exposure.

Constructing Valleys

• Flash structural members that protrude through a roof at all intersecting angles to prevent leaks (Figure 8.13).

• Most roof leaks occur where water is channeled off the roof or where the roof abuts a vertical wall or chimney. At these points, metal valleys and flashings help keep the structure sound and dry.

• Extend step flashing under the shingles or shakes and up the vertical surface. Cover it with a second layer of flashing or counterflashing.

• Different flashing metals are available for different climatic requirements.

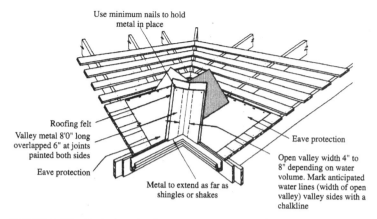

Use minimum nails to hold
metal in place

Roofing felt
Valley metal 8'0" long
overlapped 6" at joints
painted both sides

Eave protection

Metal to extend as far as
shingles or shakes

Eave protection

Open valley width 4" to
8" depending on water
volume. Mark anticipated
water lines (width of open
valley) valley sides with a
chalkline

FIGURE 8.13   Typical roof valley.

- Some metals have baked-on enamel coatings.
- Use metals that have proven their reliability under the specific conditions that are likely to be encountered once the roof is applied. Never take chances.
- Use metal flashings that have the same longevity as the shingles or shakes.
- Typical saddle flashing is shown in Figure 8.14.
- For roofs with slopes of 1/1 or greater, extend valley flashing not less than 7 inches on each side of the valley centerline.
- For roof slopes less then 1/1, extend the flashing not less than 10 inches on each side.
- Valley flashing should be center-crimped and painted, and made from galvanized steel, copper, or aluminum.
- Underlay valley metal with at least No. 15 roofing felt.
- Do not apply shingles with their grain parallel to the valley centerline.
- Cut shingles extending into the valley at the correct angle (Figure 8.15).
- Joints between shingles must not break into the valley.
- For shake roofs, follow the guidelines above for shingle valleys.
- Give the metal valley a minimum total width of 20 inches.

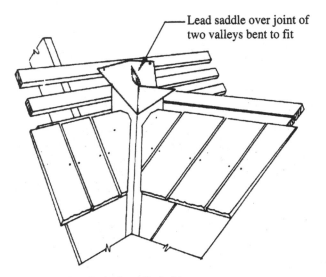

Lead saddle over joint of two valleys bent to fit

FIGURE 8.14 Typical saddle flashing.

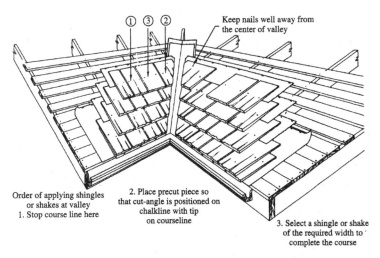

① ③ ②

Keep nails well away from the center of valley

Order of applying shingles or shakes at valley
1. Stop course line here

2. Place precut piece so that cut-angle is positioned on chalkline with tip on courseline

3. Select a shingle or shake of the required width to complete the course

FIGURE 8.15 Flashing details for shingle and shake valleys.

- In some areas, however, flashing width requirements can differ—consult local building codes.

- Recommended application details for metal flashings around typical roof projections, such as chimney and vent pipes, are shown in Figure 8.16.

- Application is accomplished in the manner described for asphalt shingles in Chapter 7.

- Apply each shingle or shake with two corrosion-resistant fasteners, such as stainless steel-type 304 or 316, hot-dipped zinc-coated, or aluminum nails.

- If preservative-treated shingles or shakes are installed, pay close attention to the preservative manufacturer's recommendations about the compatibility of the preservative chemicals with the fastener.

- Minimum nail lengths are shown in Table 8.4.

- Use aluminum or stainless steel-type 304 or 316 16 gauge staples.

- Drive two staples per shingle or shake with the staple crowns horizontal to the shingle or shake butt.

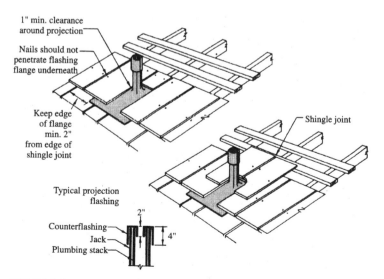

FIGURE 8.16 Flashing details for typical roof projections.

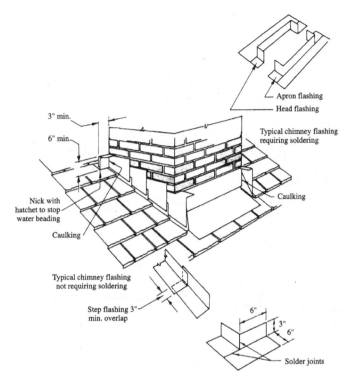

FIGURE 8.16 *(continued)*   Flashing details for typical roof projections.

- The staples should be at least ⅞₆ inch long or long enough to penetrate the sheathing by at least ½ inch.
- Drive two staples in the same location as nails relative to the sides and overlapping buttline, and flush with the surface of the shingle or shake. Preservative reaction with these metals should be checked with the manufacturer.

Detailing Vents

- Provide vents at the eaves, or soffits, as well as at gable ends.
- If cross-ventilation is desired, place vents at the ridgelines.
- Screen the vents to prevent the ingress of insects.

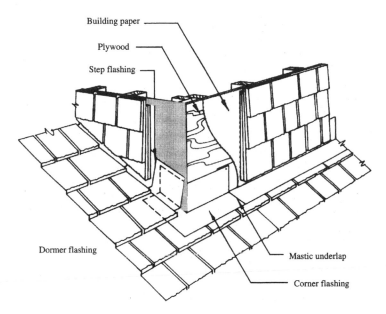

Building paper

Plywood

Step flashing

Dormer flashing

Mastic underlap

Corner flashing

FIGURE 8.16 *(continued)* Flashing details for typical roof projections.

TABLE 8.4 Fasteners

Shingle or shake type	Nail type	Min. length (in.)
Shingles—new roof		
16" and 18" shingles	3d box	$1\frac{1}{3}$
24" shingles	4d box	$1\frac{1}{2}$
Shakes—new roof		
18" straightsplit	5d box	$1\frac{3}{4}$
18" and 24" hand-split and resawn	6d box	2
24" tapersplit	5d box	$1\frac{3}{4}$
18" and 24" tapersawn	6d box	2

fastfacts

The importance of good attic ventilation beneath the roof cannot be overemphasized. Air movement prevents or inhibits moisture condensation on the undersurface of the shingles or shakes, and on the roof deck and rafters.

- For adequate ventilation, the rule of thumb is that the ratio of total net free ventilation area to the area of the attic should be not less than 1/150, with compensation made for screens over vent apertures. Attic fans can be beneficial, since these supply additional air movement in attic spaces. Several examples of construction techniques that provide roof ventilation are shown in Figure 8.17.

Designing Vapor Barriers

- The decision to use a separate vapor barrier must be made by the designer.

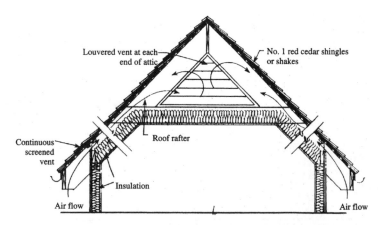

FIGURE 8.17 Examples of construction techniques for roof ventilation.

- Factors to be considered include:
  - The type of building.
  - The end use of the building.
  - The geographic location of the building.
- When a vapor barrier is used, take care to ensure that the dew point is well to the outside of the vapor barrier in order to prevent condensation on the deck.
- Ideally, the vapor barrier should be as close as possible to the warm side of the roof.
- The thickness of the insulation should be increased as the deck thickness increases to maintain the correct location of the dew point.
- A helpful product, Cedar Breather, maximizes the performance and life of cedar roofs.
- It allows for a continuous air space beneath cedar shakes and shingles when they are installed on plywood decks.
- This reduces excess moisture and minimizes the potential for shingles to rot and warp.
- Made of a patented three-dimensional nylon matrix, Cedar Breather enables the entire underside of the shingles to dry uniformly.
- Dryness prevents premature deck and shingle failure and eliminates the need for lathe or strapping.

fastfacts

➤ *In unevenly heated buildings, such as churches and halls, or buildings that generate an unusually high level of moisture, such as swimming pools, the excess humidity may have to be removed by mechanical means to prevent condensation on the roof deck.*

➤ *In air-conditioned buildings, a cold-weather roof system allows a constant flow of air between the insulation and the roofing.*

➤ *This helps reduce the energy required for cooling.*

Preparing for Reroofing Jobs

Generally, wooden shakes and shingles can be applied directly over existing roofing materials, with the exception of tile and slate. What follows are important procedures that you should use when preparing to reroof with wooden shakes and shingles (see also Figure 8.18).

- At the eaves and gable ends, cut back the old roofing material approximately 4 to 6 inches and replace it with a wooden filler of the same thickness.
- Remove the existing ridge and hip shingles before applying new roofing materials.
- When necessary, install new valleys as well as larger wall flashings, and counterflashings.
- Install a wooden strip in the existing sheet metal valleys to separate old metal from new metal.
- If an inspection of the existing roof indicates that the old wooden shingles can remain, carefully prepare the surface of the roof to receive the new material as follows.
  - Apply wooden shakes with a felt underlayment, as in new construction methods.
  - When applying wooden shingles over wooden shingles, eliminate all obstructions and voids, such as curling shingles or missing shingles.
  - Remove all loose or protruding nails and renail the shingles in a new location.
  - Nail down all loose shingles.
  - Split all badly curled or warped old shingles and nail down the segments.
  - Replace missing shingles with new ones.
  - If necessary, apply strip sheathing over existing wooden shingles.
  - When necessary, remove or cut existing siding to allow for the installation of new and larger step flashings or tin shingles.
  - Use nails that are long enough to penetrate the original roof sheathing by ½ inch.
- Figure 8.18 is handy for taking into the field when you do a reroof on a shake or shingle job.
- When the roof is subject to the impact of unusually high winds, cut back the shingles at the eaves and rakes and install 1 inch thick wooden strips that are 4 to 6 inches wide.

REROOFING JOB - PREP LIST	
	ITEM
	At the eaves and gable ends, cut back the old roofing material approximately 4 to 6 inches and replace it with a wooden filler of the same thickness.
	Remove the existing ridge and hip shingles before applying new roofing materials.
	When necessary, install new valleys, larger wall flashings, and counterflashings.
	Install a wooden strip in the exiting sheet metal valleys to separate old metal from new metal.
	IF OLD SHINGLES ARE TO BE LEFT IN PLACE
	Apply wooden shakes with a felt underlayment, as in new construction methods.
	When applying wooden shingles over wooden shingles, eliminate all obstructions and voids, such as curling shingles or missing shingles.
	Remove all loose or protruding nails and renail the shingles in a new location.
	Nail down all loose shingles.
	Split all badly curled or warped old shingles and nail down the segments.
	Replace missing shingles with new ones.
	If necessary, apply strip sheathing over existing wooden shingles.
	When necessary, remove or cut existing siding to allow for the installation of new and larger step flashings or tin shingles.
	Use nails that are long enough to penetrate the original roof sheathing by 1/2 inch.

FIGURE 8.18 Preparing for reroofing jobs.

- Nail the boards firmly and allow their outside edges to project beyond the edge of the deck by the same distance as the old wooden shingles.

- To provide a smooth deck for asphalt roofing, apply a backer board over the wooden shingles or use beveled, wooden feathering strips along the butts of each course of old shingles.

- If old asphalt shingles are to remain in place, nail down or cut away all loose, curled, or lifted shingles.

- Remove all loose and protruding nails and all badly worn edging strips.

- Replace the edging strips with new ones.

- Just before applying the new roofing, sweep the surface clear of all loose debris.

Detailing Mansard Roofs

Mansards are not as popular as typical pitched and flat roofs, but they are a design that roofing professionals work with from time to time. The downward slope of the mansard roof line visually reduces the scale of the building and offers an inexpensive method for softening a box-like appearance. Architects use this technique on large commercial projects, for example, to break up a monotonous appearance on flat-roofed apartment buildings. Properly used, a mansard roof can improve a building design without substantially increasing construction costs.

If raised up above the level of a built-up roof (BUR), the mansard can screen out roof penetrations or mechanical equipment. Two of the most widely used roofing materials on the mansard roof are cedar shingles and shakes. The light weight of shingles and shakes and the ease with which they can be applied contribute substantially to economical construction.

- The shakes or shingles can be installed over light framing, such as spaced battens.

- Though they appear to be independent of a flat roof, mansards can contribute to roof leaks.

- Make certain to work closely with the architect and the other trades to insure a watertight job.

- Always pay close attention to any head and edge flashing.

- At any joint to a parapet, make certain that the parapet cap and the head flashing of the mansard are correctly detailed and installed— this is a primary source of leaking for mansards.
- Be certain that uplift from the wind load is properly addressed.

PREPARING SPECIAL ROOF DECKS

Wooden decks are an ideal base over which to apply shingles or shakes because they can be attached in the conventional manner. The problem of how to fasten shingles or shakes is created when a layer of insulation, normally one of the rigid types, is included. Driving abnormally long nails through the shingles, the insulation, and into the deck is generally unsatisfactory.

- Horizontal strapping is required to overcome the fastening difficulties.
- Strapping requires that fewer nails penetrate through the insulation and into the deck and achieves sufficient thermal efficiency by reducing the number of conductors.
- In addition, the nail lengths can be chosen to prevent the points from protruding through the deck where they might mar the inside face.
- If ice dams are a potential problem, or if reverse condensation such as might be encountered in an ice arena is likely to occur, use the cold-weather roof system in conjunction with horizontal strapping.
- Provide ventilation at the eaves and at the peak.
- In buildings such as ski cabins, which can be subjected to heavy snow loads, fasten wooden members, typically 2 x 4 inch boards on edge, from ridge to eave on the roof deck and place rigid insulation between the members.
- Then apply strapping across the top of these members to provide a ventilated air space and avoid compressing the insulation.
- If the shingles or shakes are nailed directly through rigid insulation, a number of problems may be encountered:
  - For instance, longer nails have thicker shanks that tend to split the shingles or shakes.

fastfacts

➤ *The need for strapping can be eliminated by using a false plywood deck to which the shingles or shakes are directly fastened, immediately over the insulation.*

➤ *Exterior-grade sheathing panels are ideal for this purpose, since they provide a strong, smooth surface.*

➤ *Under certain conditions of pitch and loading, however, there might be a tendency for the entire roof above the decking to creep downward, which bends the nail fastenings, compresses the insulation, and generally reduces efficiency.*

➤ *In such cases, it is often desirable to install vertical members as previously described.*

- Movement by the shingle or shake, which is caused by natural expansion and contraction cycles due to wet and dry conditions, tends to enlarge the holes in the insulation and reduce its efficiency.

- For this reason, the use of strapping or a false plywood deck is recommended.

- Numerous types of rigid insulation are now in use.

- Rigid insulation can be made from expanded polystyrene beads, rigid urethane laminate, low-density fiberboard, or fast-setting liquids poured onsite.

- Its thickness can be more than 2 inches, and lengths and widths vary depending on the manufacturer.

- All these types of rigid insulation are efficient insulators and are usually of sufficient density to hold the weight of a normal roof covering without a lumber bridging.

- Covering metal decks with shingles or shakes presents a rather unique problem.

- These decks are often used for economic reasons, but they generally require a finish roofing capable of withstanding weather conditions.

- In addition, aesthetic considerations often require that the deck itself be covered with a material such as red cedar shingles or shakes to provide a pleasing finish.

- When a metal deck is to be covered, give consideration to the use and placement of vapor barriers.

- If insulation is to be placed on top of a metal deck, take into account the entire roofing system.

- For example, wooden members must not be sandwiched between two vapor barriers.

- If this is unavoidable, use preservative-treated wood. In some cases, the seams in a metal deck can be sealed to create an effective vapor barrier.

- This design should be passed by the design anchor or the building owner.

- Use boards or a panel deck as a nailing base for the shingles or shakes.

- Support it with vertical lumber members fixed to the deck.

- This can be achieved in a number of ways:

  - Corrugated decks. On a corrugated deck, fasten vertical lengths of lumber to the deck and horizontal boards, or apply panels across the vertical pieces. If insulation is required, place it on top of the vertical members and hold it in place with the nails fastening the boards or panels to the members (Figure 8.19).

  - Sheet decks. On a sheet deck, or where the corrugations are very shallow, it might be necessary to use angle clips to attach the vertical members to the deck. Nail the clips to the lumber and bolt or screw it to the deck. Then apply boards or panels as before.

- If there is a likelihood of excessive moisture buildup, as might be encountered in icy areas, employ the cold-weather roof principle and supplement it with a mechanically produced air flow if necessary.

DESIGNING COLD-WEATHER ROOF SYSTEMS

Cedar shingles and shakes are an excellent roofing material for cold weather areas that experience heavy snowfall and severe temperatures. A natural wood product, they offer the advantages of durability, superior wind resistance, and good thermal and acoustical

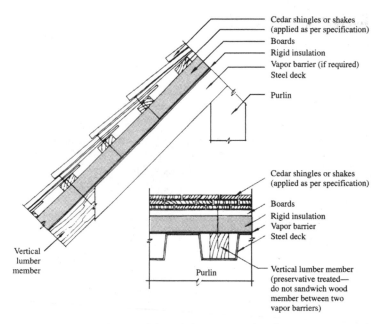

FIGURE 18.19 Cedar shingles or shakes over steel roof.

properties. As with any other roofing material, however, their best performance depends on proper design, sound construction practices, and correct installation.

- In cold-weather areas and particularly in mountainous regions that experience heavy snowfall, the cold-weather or vented roof system is recommended (Figure 8.20).
- This system allows a constant flow of cold air above the insulation, but below the roofing material.

With other roofing systems, ice buildup along the eaves can be a problem. Heat escapes from the insulation, melts the snow, and the resulting water runs down the roof to the cold overhangs, where it freezes. This can cause water to back up and penetrate the roof system. A properly installed, vented cold-weather roof eliminates this problem.

- Venting space should be sufficient to allow a free flow of air from eave to rooftop.

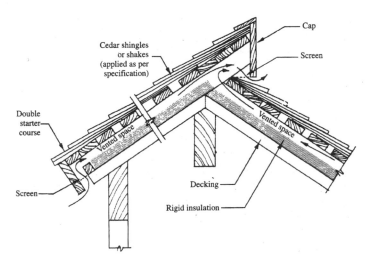

FIGURE 8.20 Details of cold-weather roof systems.

- There are a number of important considerations that influence roof performance in areas that experience heavy snowfall, particularly mountainous regions:
  - The steeper the roof, the better the performance.
  - Locate chimneys at the ridge or gable ends away from possible snow pressure on the slopes.
  - Locate plumbing pipes on inside walls, extend them between the rafters, and vent them at the ridge.
  - If this is not practical, then use galvanized iron for the plumbing vent pipes and anchor them inside the roof.
  - Plastic vent pipes that extend through the roof can be dislodged by sliding snow.
  - Avoid wide overhangs at the eaves, as they provide large cold areas for snow and ice buildups.
  - A strip of metal along the eaves helps shed ice quickly.
  - Sliding ice and snow are constant hazards and should be given primary consideration in the total building design.
  - Do not locate outside doors at the bottom of a roof slope.
  - Entrances and all pedestrian traffic areas are better situated beneath the gable ends of the roof.

- In regions of heavy snowfall, lay shingles and shakes as you would a three-ply roof.
- Use three-layer starter courses for shingled roofs.
- For shake roofs, superior construction at the eaves is achieved by using two layers of shingles and one of shakes.
- Proper nailing is very important.
- All nails should be stainless steel-type 304 or 316 hot-dipped, zinc-coated, galvanized or equivalent and placed ¾ inch from the edge and 1½ inches above the buttline.
- Take care to facilitate proper nailing when applying the sheathing boards.
- Increase shingle or shake sidelap to 2 inches.
- Lay the entire roof with the same precautions you would take with any other type of wooden shingle or shake roof.
- Include eave protection and an interlay felt between the shakes.
- Properly installed and designed cold-weather roofs result in a sound roof system that gives many years of service during severe extremes of winter temperatures and snowfall.

APPLYING UNIQUE ROOF PATTERNS

One of the most unusual applications of cedar roofing ever used in this country is found in an architectural style that flourished in the 1920s. The roofs of this style, known curiously as English cottage thatch, thatch effect, and shingle thatch, reflect 16th century rural England and Ireland, where thatched-roof peasant cottages were commonplace in the verdant countryside.

Rolled Eaves and Gables

Possibly the most striking characteristics of the country-cottage roof are its rolled eaves and rolled gables.

- These rolled edges can be achieved through pre-assembled fascia boards with attached radius blocking.
- They are usually sold in 10 foot sections and are nailed to existing roof framing members the same way a conventional fascia board is installed.

- Any method of gable construction is acceptable because the pre-assembled gable sections are nailed on like a conventional fascia board.

- The radius of the blocking is designed to work with preformed shingles.

- The No. 1 cedar shingles come preformed and precut for installation.

- A designed framing-component system has precut and some pre-assembled pieces.

- This enables builders basically to frame roofs with conventional procedures and then attach the manufactured component system to the existing framework.

- The pre-bent shingles, which are used on all rolled sections, are of two different types:

  - Eave-bent shingles are bent across the grain of the wood.

  - Gable-bent shingles are bent with the grain of the wood.

- Both types of preformed shingles are available with either a straight buttline or a thatched buttline.

- As the name implies, the eave-bent shingles, which are bent at approximately a 20 inch radius, are used from the rolled eaves to where the roll merges with the roofline.

- The starting course at the eave is doubled.

- Either straight-butt or thatched-butt shingles can be used for the starter course.

- Eave-bent shingles also are used at the start of eyebrow dormers and rolled roof vents.

- Gable-bent shingles are used on the gables and can be bent to a 10 inch radius.

- They are more or less flexible and can be used on almost any radius.

- Gable-bent shingles can also be used on hips and in valleys.

- Enough mildly flexible shingles usually can be found in the bundles of straight shingles, however, to cover these areas.

- Gable-bent shingles come slightly over-bent so that the piece can be pressed into place, which causes the shingle to hug the contour of the roof more tightly.

- On the rounded surface of the eaves it might be necessary to occasionally nail down the butt end of a shingle.

fastfacts

Next to the rolled eaves and gables, there is no other feature of the country cottage more integral to the overall beauty of the finished product than the wave coursing. The artistic textures of the various styles of wave coursing create a dimension that cannot be found on any other type of roof.

Wave Coursing

- Wave coursing is achieved by using a thatched-butt shingle.
- Wave coursing fits basically into two design categories.
- Patterned designs are shown in Figure 8.21 and random designs are shown in Figure 8.22.
- In either style, the waves are achieved in a similar fashion. The long side of each shingle is laid on the short side of the previous one until the maximum exposure is reached.
- To bring the course down again to the minimum exposure, the procedure is reversed.
- The short side of each shingle is laid on the long side of the previous one.

Graduated Exposure

- Among the many variations that can be achieved with hand-split shakes is one that is produced by reducing the exposure of each course from eaves to ridge. This requires shakes of several lengths.
- A very serviceable graduated exposure roof can be built by starting at the eaves with 24 inch shakes, laid 10 inches to the weather.
- Lay one-half of each roof area with shakes of this length.
- Reduce the exposure gradually to about 8½ inches.
- Complete the area with 18 inch shakes, starting with an 8 inch exposure and diminishing to a 5 inch exposure.
- The resulting appearance exaggerates the actual distance from eaves to ridge.

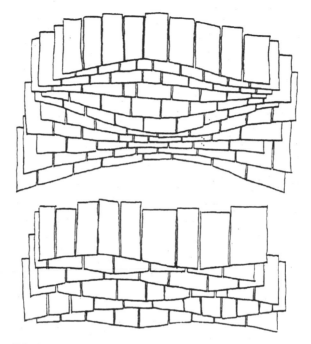

FIGURE 8.21 Pattern designs.

Tilted Strip at Gables

While shake roofs are less prone to drip rainfall or snowmelt from the gables than are roofs of smoother materials, gable-drip and icicles can be eliminated by inserting a single strip of beveled cedar siding the full length of each gable end with the thick edge flush with the sheathing edge. The resulting inward pitch of the roof surface keeps moisture away from the gable edge. It is also a pleasing accent to the gable line.

Staggered Lines

To secure irregular and random roof patterns, shakes can be laid with butts placed slightly below or above the horizontal lines that govern each course. If an extremely irregular pattern is desired, longer shakes can be interspersed in the roof with their butts several

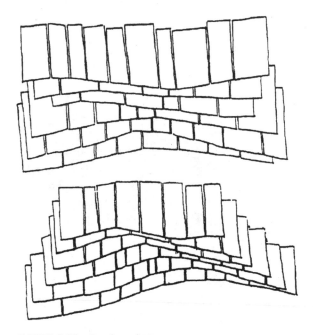

FIGURE 8.22 Random designs.

inches lower than the course lines. Mixtures of very rough shakes and fairly smooth shakes or wooden shingles can produce a more rugged appearance than a roof composed entirely of rough shakes.

TREATING WOODEN ROOFS

Weathering

To prevent owner dissatisfaction and complaints, it is in the best interest of the roofing contractor to explain to the customer the effects of weathering on a shake or shingle roof before contracts are signed. The surface of any untreated wood product changes when exposed to the effects of sunlight and precipitation. This change is partly physical and partly chemical. It is cumulatively referred to as weathering.

- The first noticeable change is the color.
- The initial red-brown color tends to fade, and a graying process begins.
- This change to a silvery gray results when ultraviolet radiation from the sun strips the surface layer (less than 0.01 inch deep) of certain cell-wall materials.
- This first change in color is rather rapid, occurring within the first year of exposure or, under more severe conditions, within several months.
- Gradually, the silver gray changes to a darker, more graphite gray, which indicates the colonization of micro fungi on the surface and completes the initial phase of the weathering process.
- Good housekeeping for a wooden roof requires the removal of all leaf litter, pine needles, and debris that accumulate over time between the shingles and shakes and in the valleys of the actual roof structure.
- It is best to do this housekeeping before the wet season begins.
- In the Pacific Northwest, this is before the autumn rains begin. Along the Gulf Coast or in areas that are conducive to the growth of molds, mildews, and fungi, this task might be required several times each year.
- The accumulated debris has two detrimental effects on the roof:
  - First and foremost, debris tends to retard the shedding of water.
  - Second, and more destructively, debris retains moisture that allows wood destroying fungi to grow. Wood-rotting fungi require water as part of their life cycle.
- If the amount of water on roofs is reduced so that the minimum level is never reached, mold and fungus organisms cannot grow.
- To promote drying, it is important to remove overhanging tree branches that excessively shade roof sections and retard the drying of the roof surface.
- Never allow tree branches to touch and rub against the surface of the roof.
- The mechanical action of this rubbing can literally wear sweeping grooves into the surface of the shingles or loosen the fasteners that hold them.
- Finally, in excessively woody areas, branches can retard the flow of air so that surface drying occurs very slowly.

- In all cases, use sound judgment about pruning, trimming, or removing problem-causing trees.

- Excessive growth of lichens and mosses is a sure indicator that a roof is not drying properly and that enough moisture is available to support wood-rotting organisms.

Roof Cleaning

Removing the accumulated debris is relatively simple. In most cases, anyone who is willing to climb on the roof can wash off most of the material with a garden hose. A stiff broom is also handy for removing all the leaf and pine-needle litter. It is easy to remove litter from around chimneys and in valleys, but all those areas between the individual shingles and shakes, or keyways, can be more difficult to clean. Wet sweeping the surface in a careful manner also removes the initial growths of moss and lichens.

Professional services sometimes use high-pressure washers. These are effective, but they must be used with extreme care because cedar is a soft, low-density wood. Excessive or imprudent use of high-pressure systems can detach shingles or inflict many years of wear in moments, even though the roof initially looks like new.

- Most cedar roofs remain untreated and are known to provide excellent service.

- A surface chemical treatment is desirable, however, under conditions that can be conducive to premature deterioration, such as areas whose climate combines heat and humidity for considerable portions of the year, roofs of low pitch or slope, roofs beneath overhanging trees, etc.

- There are naturally resilient oils present in cedar, so it is more advantageous to let the roof weather for approximately one year before applying a treatment. Allowing the oils to leach out gives the wood a greater porosity to absorb the preservatives.

Applying Chemicals

- Fungicidal chemicals inhibit moss, fungus, and mildew and contribute to a roof's life service.

- When applying one of these chemicals, keep in mind that they are more effective when the moss is actively growing than when it is in the dormant phase. In most parts of our country, moss grows vigorously during rainy seasons.

- If chemicals are applied during the dry season, it is recommended that you wet the moss first to promote chemical uptake.

- To control runoff and increase effectiveness, apply when rain is not likely.

- It may be several days before the growth dies and can be effectively removed.

- Remember to use care when handling and disposing of chemically treated moss and litter to reduce accidental exposure.

- The corrosive nature of these chemicals means that unnecessary damage must be avoided by thoroughly rinsing metal tools, gutters, and flashings that come in contact with the solutions.

- Exercise care when applying some chemicals due to their toxicity. Follow the manufacturer's directions.

Preventing Excessive Dryness

In areas of low humidity, where moss, fungus, and mildew are not a potential problem, commercial oil-based preservatives can prevent excessive dryness and prolong the natural resilience of the cedar. In such areas, fungicidal chemicals can be used to good effect, but are not required. Remember that these surface treatments are temporary and should be applied approximately every five years, depending on the products used. When applying preservative chemicals to shakes and shingles, consider several important issues:

- First, the coating should be as uniform as possible.
  - Avoid drips and runs.
  - It is better to apply several light coats and obtain good absorption than to try to force the chemicals all in at once.
  - The important point is to reach a certain level of preservative retention.
  - A low retention level results in an ineffective treatment.
- Second, since much of the decay occurs in the butt region, be conscientious when applying the solution so that the exposed butts are covered as well as the exposed shingle surface.
- The three methods available to provide proper application are brushing, using a thick-napped roller, or spraying.

- Because of the irregular surface, a multi-gallon pump-type sprayer, set to produce a coarse spray, results in a more uniform coating.

- Spraying requires a calm day, with a wind less than 4 miles per hour, to prevent overspray from drifting over plants or into neighboring yards.

- Brushes are also effective because they allow for the thorough treatment of the keyways and ends.

- The Cedar Shake and Shingle Bureau recommends CCA pressure treated cedar shakes and shingles in Southeastern states with a climate index of 65 or greater as identified by the U.S. Department of Agriculture's Forest Service.

- The pressure treatment is permanent and extends the service life of the roof.

- Most CCA-treated shingles and shakes are available with a 30-year warranty.

- The services referred to in the paragraphs above are very important and can be translated into an excellent cash flow.

- Set up service contract sales while developing the original roofing contract.

- By doing so, you are working smart rather than hard—being paid to monitor your warranties.

SLATE ROOFS

The principal difference between slate and other stones is the natural cleavage in the slate. This cleavage permits it to be split, forming sheets of various thicknesses. A slate roof is typically a custom product and its use gives character to a building unlike that of any other covering. Slate quarried for roofing is of dense, sound rock and is exceedingly tough and durable.

- Roofing slate comes in thicknesses from ¼ to 1 inch and offers roof appearances that range from smooth and uniform to rough and textured.
- Slate roofs have a service life of 75 to 100 years.

EXAMINING SLATE CHARACTERISTICS

As previously mentioned, slate has a natural cleavage that permits it to be more easily split in one direction than in others. A second direction of fracture or scallop, which is usually at right angles to the cleavage, is called the grain. Roofing slates are commonly split so that the length of the slate runs with the grain. Slate, like any other stone, becomes harder and tougher upon exposure than it is when first quarried—it is practically nonabsorptive.

- Tests show that typical slate has a porosity of 0.15 to 0.4 percent.
- Many slates split to a smooth, practically even and uniform surface, while others are somewhat rough and uneven.
- As a result, a wide range of surface effects is available for the finished roof.
- Slate from certain localities contains comparatively narrow bands of rock differing to various degrees in chemical composition and color from the main body of the stone.
- These bands are called ribbons.
- Ribbons that do not contain injurious constituents and are of desirable color are not objectionable.
- Slates of this type, when trimmed so that the ribbons are eliminated, are known as clear slate.
- Slates that contain some ribbons are sold as ribbon stock.

Picking Colors

- The color of slate is determined by its chemical and mineral composition.
- Since these factors differ in various localities, it is possible to obtain roofing slates in a variety of colors and shades.
- It is truly remarkable to find a natural product that possesses, in addition to its other qualities, such unlimited color possibilities.
- Surface colors can be uniform, or have contrasting hues.
- Moreover, if the design of the building requires a roof of one general color, it can be graded up or down the slope from dark to light as desired.

fastfacts

For the purpose of classifying the basic natural colors of roofing slate now available in large quantities for general use, the Division of Simplified Practice of the U.S. Department of Commerce recommends the color nomenclature for slate materials shown in Table 9.1.

TABLE 9.1 Basic Slate Colors

Black	Grey	Purple	Green
Blue black	Blue grey	Mottled purple and green	Red

- To liven the monotony of a flat, uniform body color, various shades of the same color can be used to provide interesting variations up and down, across, or interspersed throughout the roof.

- A low eaved, prominent roof surface in a quiet or contrasting blend of autumnal colors can cause the structure to blend with its surroundings.

- Slate color can be unfading, or if weathering slate is used, the roof color can mellow with age and weather.

- These extremes and all the steps between are available to roof designers.

- To take advantage of the various available effects, the source of the slate, as well as its ultimate color characteristics, should be known.

- For instance, Pennsylvania slate colors are blue-gray, blue-black, and black.

- Buckingham and other Virginia slates are generally blue-gray to dark gray with micaceous spots on the surface that produce an unusual luster.

- Vermont slates can be light gray, black-black, unfading and weathering green, unfading purple, and variegated mottled purple and green.

- Unfading red slates are found only in Washington County, New York. These are the most costly.

Estimating Quantities

- In the United States, slate is sold by the square.

- A square of roofing slate is defined by the U.S. Department of Commerce, Bureau of Standards, in Simplified Practice Recommendation No. 14, as follows:

- A square of roofing slate means a sufficient number of slate shingles of any size to cover 100 square feet of plain roofing surface, when laid with approved or customary standard lap of 3 inches.

- Slates for surfacing flat roofs are usually laid tile fashion, without lap, in which case a square of slate would cover an area greater than 100 square feet.

- The quantity per square varies from 686 pieces for the 10 x 6 inch size to 98 for the 24 x 14 inch size, which includes the allowance for a 3 inch headlap.

- Note that for roofs of comparatively little slope, where a 4 inch lap is required, an additional quantity must be provided.

- For steep roofs or siding, where a 2 inch lap is sufficient, fewer slates are necessary.

- However, slate is always sold on the basis of quantity required for a 3 inch lap even for flat roofs.

- Table 9.2 shows the sizes for standard ³⁄₁₆ inch thick slate, the minimum number of slates required per square, the respective exposures for the slates listed, and the weight (per square) of the nails used to secure each size of slate.

- The following is a guide to the many factors that should be taken into consideration when beginning a slate roof project.

- For rough estimates, a good slate roof costs from 6 to 8 percent of the total cost of the average building or home.

- From the time work is first started at the quarry until the material is laid on the building, there are certain costs that must be taken into consideration (Figure 9.1) including:

  - Cost of slate (punched) on cars at the quarry.

  - Freight from quarry to destination.

fastfacts

➤ *Each roofing contractor has a method of compensating for waste, breakage, projections through roofs, dormers, hips, ridges, valleys, and other factors occasionally encountered in roofing work.*

➤ *The method is usually based on experience, labor skill, local conditions or practice, and items peculiar to a locality.*

TABLE 9.2 Schedule for Standard ⅗₆ Inch Thick Slate

Size of slate (in.)	Slates per square	Exposure with 3" lap (in.)	Nails per square (lb, oz)	
26 × 14	89	11½	1	0
24 × 16	86	10½	1	0
24 × 14	98	10½	1	2
24 × 13	106	10½	1	3
24 × 11	125	10½	1	7
24 × 12	114	10½	1	5
22 × 14	108	9½	1	4
22 × 13	117	9½	1	5
22 × 12	126	9½	1	7
22 × 11	138	9½	1	9
22 × 10	152	9½	1	12
20 × 14	121	8½	1	6
20 × 13	132	8½	1	8
20 × 12	141	8½	1	10
20 × 11	154	8½	1	12
20 × 10	170	8½	1	15
20 × 9	189	8½	2	3
18 × 14	137	7½	1	9
18 × 13	148	7½	1	11
18 × 12	160	7½	1	13
18 × 11	175	7½	2	0
18 × 10	192	7½	2	3
18 × 9	213	7½	2	7

TABLE 9.2 *(continued)*   Schedule for Standard ³⁄₁₆ Inch Thick Slate

Size of slate (in.)	Slates per square	Exposure with 3" lap (in.)	Nails per square (lb, oz)	
16 × 14	160	6½	1	13
16 × 12	184	6½	2	2
16 × 11	201	6½	2	5
16 × 10	222	6½	2	8
16 × 9	246	6½	2	13
16 × 8	277	6½	3	2
14 × 12	218	5½	2	8
14 × 11	238	5½	2	11
14 × 10	261	5½	3	3
14 × 9	291	5½	3	5
14 × 8	327	5½	3	12
14 × 7	374	5½	4	4
12 × 10	320	4½	3	10
12 × 9	355	4½	4	1
12 × 8	400	4½	4	9
12 × 7	457	4½	5	3
12 × 6	533	4½	6	1
11 × 8	450	4	5	2
11 × 7	515	4	5	14
10 × 8	515	3½	5	14
10 × 7	588	3½	7	4
10 × 6	686	3½	7	13

- Loading and hauling to storage yard.
- Unloading, piling, and waste at storage yard.
- Loading and hauling to job.
- Unloading and piling at job.
- Placing on roof and laying.
- Roofing felt.
- Nails
- Elastic or plastic cement.
- Snow guard or snow rails.
- Sheet metal.
- Labor, including compensation insurance.
- Waste in handling, cutting, and fitting.
- Contractor's overhead on organization and equipment.
- Cost of guarantee or bond.
- Contractor's profit.

The list in Figure 9.1 is a handy version of the list above. It is useful to print for the field. While it might seem a comparatively simple problem to estimate the net quantity, it is not as easy to allow for the additional material required for slate around chimneys, dormers, hips, valleys, etc. These allowances depend largely on the judgment and experience of the estimator and the design of the roof.

Estimating Weights

- The weight of a square of slate varies from 650 to 8000 pounds, depending on the thickness of each slate.
- Slates of commercial standard thickness (approximately $\frac{3}{16}$ inch) weigh from 650 to 750 pounds per square.
- The dead-load weight of slate per square, however, can be estimated at a maximum of 800 pounds, or 8 pounds per square foot, when the weight of the slate, felt, and nails is combined.
- The weight of slate shingles depends on the size, the color, and the source quarry.
- Table 9.3 shows the weight per square of slate products of different thicknesses for both sloped and flat roofs.
- The actual weight of slate products can vary from 10 percent above to 15 percent below the weights shown in the table.

	COST CONSIDERATIONS FOR SLATE ROOFS
	ITEM
	Cost of slate (punched) on cars at the quarry.
	Freight from quarry to destination.
	Loading and hauling to storage yard.
	Unloading, piling and waste at storage yard.
	Loading and hauling to job.
	Unloading and piling at job.
	Placing on roof and laying.
	Roofing felt.
	Nails
	Elastic or plastic cement.
	Snow guard or snow rails.
	Sheet metal.
	Labor, including worker's comp insurance.
	Waste in handling, cutting and fitting.
	Contractor's overhead on organization and equipment.
	Cost of guarantee or bond.
	Contractor's profit.

FIGURE 9.1   Choosing slate.

INSTALLING SLATE ROOFS

As with all other roof installation methods, there are certain important considerations that must be kept in mind before you start to lay a slate roof.

Using Slater's Tools

- The tools commonly used by the slater are:
  - Punch
  - Hammer
  - Ripper
  - Stake
- If much punching is done at the yard or on the job, use a punching machine to punch the nail holes and cut the slates.
- It is adjustable to any size or shape, and it cuts and punches a countersunk hole in one operation.

TABLE 9.3 Average Weight of Slate Per Square

Slate thickness (in.)	Sloping roof with 3" lap (lb per square)	Flat roof without lap (lb per square)
$3/16$	700	240
$3/16$	750	250
$1/4$	1000	335
$3/8$	1500	500
$1/2$	2000	675
$3/4$	3000	1000
1	4000	1330
$1 1/4$	5000	1670
$1 1/2$	6000	2000
$1 3/4$	7000	
2	8000	

- The hand or mawl punch is forged from fine tool steel that has been hardened and ground.
- It is about 4½ inches long with one tapered end.
- The butt end is struck with a mawl to punch the nail hole.
- Slater's tools are all dropforged.
- An approved hammer is forged solid, all in one piece, from crucible cast steel, with an unbreakable leather handle to avoid slipping and blistering the hands.
- One end terminates in a sharp point for punching slate, the other in the hammerhead.
- There is a claw in the center for drawing nails, and on each side of the shank there is a shear edge for cutting slate.
- The head, point, and cutting edges are tempered to withstand heavy work.
- The slater's stock size hammer has a 12 inch handle.

- The ripper is about 24 inches long and is forged from crucible cast steel.
- It is used to remove broken slate and make repairs.
- A hook on the end can be used to cut and remove slating nails.
- The blade is drawn very thin, and the hook end is correctly tempered for hard wear.
- The stake is about 18 inches long and T-shaped.
- The long edge is used as a rest upon which to cut and punch slate, or as a straight edge to mark the slate when cutting and fitting around chimneys, hips, valleys, etc.
- The short arm is tapered and pointed for driving into a plank or scaffold.
- These tools, as well as 24 inch stakes, shorter rippers, left-handed hammers, special hammers, and other tools, can be obtained from any slate producer or tool manufacturer.
- A nail pouch, tinner's snips, rule, and chalkline complete the slater's equipment.

Laying Underlayment Felts

A watertight, standard slate roof can be laid on open lath boards instead of on felt, as is often done on buildings that do not require heat. If felt is used, the thickness of the felt has a relatively minor effect on the watertight integrity of the roof. If felt is installed as soon as the roof is sheathed, it protects the building from moisture until the slates are laid. Felt also forms a cushion for the slate. This cushion is especially important when thick slates are used.

- For standard slate roofs, apply a No. 30 felt.
- For graduated slate roofs, apply two layers of felt and stagger the felt joints and laps.
- As with tile, whenever slate is laid on a roof with a minimum slope of 4 inches per foot or less, a double layer of felt set in either mastic or hot asphalt is recommended to ensure a watertight job.
- Felts for use in slate roofing should be asphalt-saturated felts and No. 30 or heavier.
- Guidelines for applying felts for slate roofing purposes:
  - Lay felts in horizontal layers with joints lapped toward the eaves and at the ends.

- Employ a lap of at least 3 inches, and secure the edges of the felts to the surface over which the felts are laid.

- Extend the felt over all hips and ridges at least 12 inches to form a double thickness.

- Employ a lap of not less than 2 inches over the metal lining of valleys and gutters.

- All felt should be preserved unbroken, tight, and whole.

- Many roofers drive the slater's stake into the roof boards. To avoid damaging the roofing felt, use a plank for this purpose or drive the stake into the scaffold only.

- In locations where the January mean temperature is 30°F or lower, apply two plies of No. 15 felt or one ply of No. 50 felt as the underlayment to serve as an ice shield.

- Work from the eaves to a point 24 inches inside the inside wall line of the building.

- Set the felt in hot asphalt or mastic, or an adhered bitumen membrane.

Nailing Slate

- Like any other construction unit, a slate roof can only be as strong and enduring as its weakest part.

- Most slate roof failures can be attributed to the punching of the nail holes, the nailing of the slates, or the nails themselves.

- Each piece of slate should have at least two nail holes.

- The standard practice is to machine punch, at the quarry, two holes in all architectural roofing slate ¼ inch and thicker.

- Four holes should be used for slates that are more than 20 inches in length and ¾ inch or more in thickness.

- Holes are punched ¼ to ⅓ the length of the slate from the upper end, and 1¼ to 2 inches from the edge (Figure 9.2).

- When four holes are used, it is customary to locate the two additional holes about 20 inches above the regular holes.

- The heads of slating nails should just touch the slate.

- Do not drive the slate "home" or draw the slate.

- The nail head should clear the slate so that the slate hangs on the nail.

fastfacts

➤ *Machine punching is preferable to hand punching.*

➤ *The term hand punching usually refers to the use of the double-headed slater's hammer.*

➤ *Machine punching can be done either at the quarry or on the job.*

➤ *Hand punching is necessary to fit hips and other complete joints.*

- Slate that is held too rigidly in place can shatter around the nail hole and eventually ride up and over the nail and be blown off in a heavy wind.
- The blame for losing tiles when they are nailed too tight is often placed on the material, whereas the real reason is the nailing method (Figure 9.2).
- All nails should penetrate the sheathing and not the joints between boards.
- This is especially important near the ridge of the roof.
- Slating nails should be 10 gauge, large-head hard copper wire nails, cut brass, or cut yellow metal type.
- Standard thickness slate usually can be applied with 1½ inch nails.
- Use 2 inch nails for hips and ridges.
- Thicker slate requires longer and heavier gauge nails.
- For all practical purposes, the ordinary diamond point and smooth shaft nail are sufficient.
- The needlepoint nail is seldom necessary or advantageous.
- The nail shaft, since it supports a greater weight and must resist a small shearing stress, should be larger than that of the shingle nail.
- To prevent the slate from being lifted up and over the nail after it is laid, the diameter of the nail head should be greater than that of shingle nails.

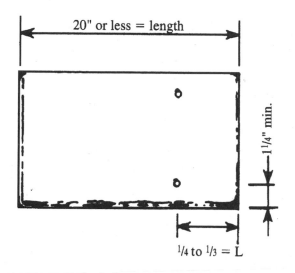

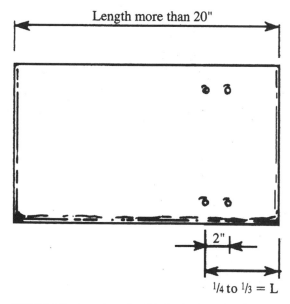

FIGURE 9.2 Location of nail holes in slate shingles.

Applying Elastic or Plastic Cement

- Elastic or plastic cements are used under slates at hip and ridge locations to help secure those slates that are smaller than regular roofing slates and that cannot be nailed easily.

- Plastic cement is also used for pointing the peaks of hips and ridges.

- Elastic or plastic cement must be waterproof.

- It must have a high melting point to prevent the slates from slipping under the heat of the sun, and a low freezing point so that it does not become brittle and crack in cold weather.

- The cement should be oily and sticky so that it adheres thoroughly to the roof.

Handling Materials

- When piling slate, pay attention to the foundations of each pile, how you start the piles, how you arrange the piles and individual slates, and how you separate the tiers.

- The tiers of slate can be kept level only if the foundation is level and free from settlement.

- The earthen foundation, upon which the slates are to be piled, should be level, dry, and solid.

- A layer of 2 inch thick wooden plank material keeps the slate off the ground, distributes the load, and helps maintain straight and even piles.

- When stacking slate, start the first tier by stacking one pile of slate flat and to a height equal to the width of the slate.

- For example, when 20 x 12 inch slate is used, the flat pile of the first tier should be 12 inches high.

- The slates piled on top of the first tier should be placed in an upright position, on edge lengthwise, and kept as straight and vertical as possible.

- During inclement weather, cover piled slate with a tarp.

DESIGNING A SLATE ROOF

Slate roofs are typically classified into three architectural designs—standard, textured, and graduated slate roofs. These are the typical slate roof classifications; however, we have listed others for the roofer's convenience.

Standard Slate Roofs

- Standard slate roofs are made with standard commercial slate that is approximately $\frac{1}{16}$ inch thick and of a standard length and width, with square tails or butts laid to a line.
- Slate of this type is commonly obtainable in the basic slate colors.
- Standard roofs are suitable for any building design that calls for permanent roofing material at a minimum cost.
- It differs from other slate roofs only in characteristics that affect the texture or appearance of the roof, such as the shape and thickness of the individual pieces.
- If desired, the butts or corners can be trimmed to give a hexagonal, diamond, or gothic pattern for all or part of the roof, for instance, as on a church spire.
- Standard roofs are sometimes varied by laying two or more sizes (lengths and widths) of standard commercial slate on the same area.

Textured Slate Roofs

- The term textured is used to designate those slates usually of rougher texture, with uneven tails or butts, and with variations in thickness or size (Figure 9.3).
- In general, this term is not applied to slate more than 3 inches thick.
- Varying shades are frequently used to enhance the color effect, which, with the characteristics just mentioned, adds interest in line and texture to the roof design.
- In addition to the basic colors of the commercial grades, accidental colorings of bronze, orange, etc., can also be used in limited quantities.

Graduated Slate Roofs

- The graduated roof combines the artistic features of the textured slate roof with additional variations in thickness, size, and exposure.

FIGURE 9.3 Textured state roof.

- The slates are arranged so that the thickest and longest occur at the eaves.

- The pieces then gradually diminish in size and thickness to the ridges.

- Slates for roofs of this type can be requested in any combination of thicknesses from ³⁄₁₆ to 1½ inches and heavier.

- The graduated slate roof presents many opportunities for variation and offers excellent possibilities for interesting treatments.

- It is referred to as a custom roof because it often is designed to harmonize with the general character of the building.

- Many producers and distributors maintain special design staffs to help architects secure the most suitable and satisfactory graduated slate roofs.

- These services are freely offered to designers and architects in the interest of a better and more harmonious slate roof.

Other Types of Slate Roofs:

- Dutch Lap is placed with regular slate on shingle lath or tight sheathing (Figure 9.4).

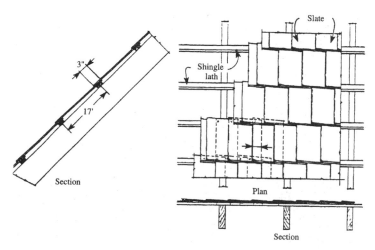

FIGURE 9.4 Dutch lap.

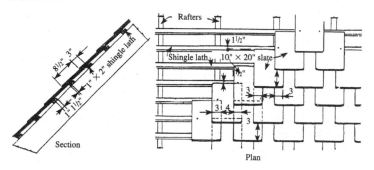

FIGURE 9.5 Open slating.

- Open slating is especially suitable for barns and other buildings, or where ventilation is desirable (Figure 9.5).
- French diagonal (or Hexagon) slating method is less expensive than the standard American method (Figure 9.6).
- When using this method to lay new roofs, if heat conservation is required, apply 30 pound (instead of 14 pound) asphalt saturated felt on a solid roof deck of sheathing or roof boards.
- This is a popular method for both new construction and reroofing jobs in New England.

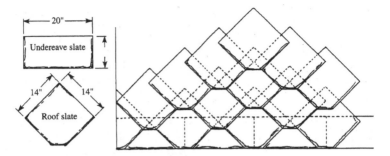

FIGURE 9.6 French method.

LAYING SLATES

Workmanship is as essential as the proper selection of the material. The more enduring the material, the more important skilled labor becomes. With slate, the most lasting roofing material known, it is important that application guidelines be followed.

- Slate application should be handled by a knowledgeable contractor who employs well trained, experienced workers.
- Through joints from the roof surface to the felt should not occur.
- Separate the joints in each slate course from those below. Otherwise, water can migrate through the joints and cause felts to disintegrate and leaks to develop in the roof.
- When slates of random widths are used, the overlapping slate should be joined as near the center of the underlying slate as possible and not less than 3 inches from any underlying joint.
- When all slates are of one width, start every other course with a half slate or, where available and practical, with a slate that is one-and-a-half times the width of the other slates, as shown in Figure 9.7.

Determining Slate Exposures

- The exposure of a slate is the portion not covered by the next course of slate above.
- This portion is the length of the unit exposed to the weather.

- The standard lap of the alternate courses used on sloping roofs is 3 inches; this is the basis on which all roofing slate is sold and the quantity computed.

- To find the proper exposure for individual jobs, deduct 3 inches from the length of the slate and then divide by 2.

- For instance, the exposure for a 24 inch slate is 24 inches minus 3 inches equals 21 inches. Divide the total by 2 and the result is an exposure of 10½ inches.

- Use Table 9.4 to obtain the proper exposure.

- Construct sloping roofs that have a rise of 8 to 20 inches per foot of horizontal run with the 3 inch lap.

- For steeper roofs, such as the mansard and others that are nearly vertical in plane, a 2 inch lap is usually sufficient.

- In some sections of the country, it is customary to increase the lap to 4 inches when the slope is from 4 to 8 inches per foot, while in other parts the 3 inch lap is considered entirely adequate.

- Employing a headlap of less than 3 inches reduces the amount of material over which water can be blown and increases the possibility of leaks.

Finishing Ridges

There are two common methods for finishing the roof ridge. These are usually known as the saddle ridge and combing ridge, but each might have other names and different laying procedures according to local practice.

fastfacts

Buildings located in the southernmost part of the country or on the Pacific slope, can be safely roofed with a lap of 2 inches provided that a high standard of workmanship is maintained.

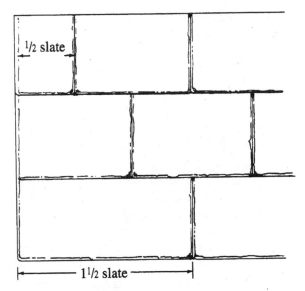

FIGURE 9.7 Starting slate.

TABLE 9.4 Exposure for Sloping Roofs

Length of slate (in.)	Exposure (in.)*
24	10½
22	9½
20	8½
18	7½
16	6½
14	5½
12	4½
10	3½

*Slope 8 to 20 inches per foot, 3-inch lap.

Saddle Ridges

- For the saddle ridge method, extend regular slates to the ridge so that pieces of slate on the opposite side of the roof butt flush.

- On top of the last regular course of roofing slate, lay another course of slate (called *combing slate*), and butt flush pieces of slate on the opposite sides of the roof.

- Lay the combing slate with the grain horizontal.

- Use a width that maintains an approximately uniform exposure or gauge.

- For example, if 20 x 12 inch slates are applied with an 8½ inch exposure, lay 12 x 8 inch slates horizontally on the ridge.

- Note that the combing slates should overlap and break joints with the underlying slate.

- In this way, all nails in the combing slate are covered by the succeeding slates, except for the nails in the last slate, which is called the finishing slate.

- Cover the nails in the finishing slate with plastic or elastic cement.

- Fill the joints on top of the ridge formed by the butted edges of the combing slate with cement when these joints are subject to heavy rainfall (Figure 9.8).

- Many architects prefer to keep the grain of the slate vertical, using combing slate that is the same width and exposure as the regular slate used on the roof. In such cases, the starting slate should be a slate-and-a-half wide rather than a half-slate wide.

- A variation of the saddle ridge is known as the strip-saddle ridge (Figure 9.9).

- Lay this type of ridge in a manner similar to the saddle ridge, but do not overlap the combing slates.

- Rather, the combing slates should butt flush and be attached with four nails.

- The combing slate for strip-saddle ridges can be the same width as the regular slate used on the roof, or narrower if the designer wishes.

- Cover the four nails with plastic cement, and set the edges of the combing slate in plastic or elastic cement.

- Clarify all design criteria with care and include them in contract documents.

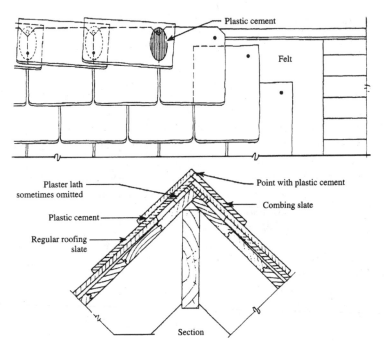

FIGURE 9.8 Saddle ridge.

Combing Ridges

- The combing ridge is laid in the same manner as the saddle ridge, except that the combing slate of the north or east side extends beyond the ridgeline, as shown in Figure 9.10.

- This extension should not be more than 1 inch.

- This type of ridge can be laid with the combing slate grain either vertical or horizontal.

- In either case, the edge of the slate should be set in elastic cement and the nails covered with elastic cement.

- If the top or combing course projects ⅟₁₆ to ⅛ inch above the top courses, it makes a better finish and is more easily filled with elastic cement.

- A variation of this type of ridge is known as the coxcomb ridge. The combing slate alternately projects on either side of the ridge.

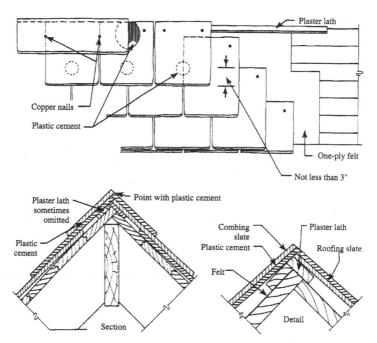

FIGURE 9.9 Strip-saddle ridge.

Finishing Hips

There are several ways to form hips on slate roofs. Descriptions of the most common follow.

Saddle Hips

- The saddle hip can be formed by placing one or two plaster laths or a 3½ inch cant strip on the sheathing.
- This forms the hip.
- Run the roofing slate up to this strip.
- On top of the cant strip and the slate, lay the hip slates, which are usually the same width as the exposure of the slates on the roof, although they can vary in width on different classes of work.
- Drive the four nails used to fasten the hip slate to the roof into the cant strip.

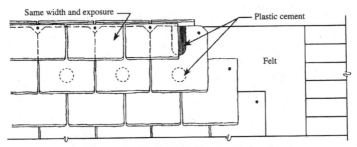

Combing slate laid with grain vertical

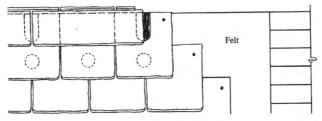

Combing slate laid with grain horizontal. Smaller
slate of proper size can be used to give same
exposure as rest of roof courses.

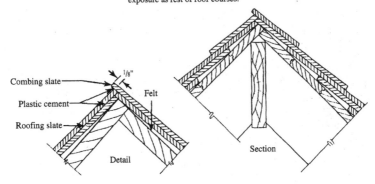

FIGURE 9.10 Combing ridge.

- Do not nail between the joints of the slate.
- Cover the heads of these nails with elastic cement and embed the lower part of the next slate.
- Elastic or plastic cement is also placed on the joint between the roofing slate and the plaster lath and on the peak of the hip before the hip slates are laid.
- A variation of the saddle hip is known as the strip-saddle hip.
- This method is used on less expensive work and can be made with narrower slates laid with butt joints that do not necessarily line up with the course of the slate on the roof.

Mitred Hips

- For this type of hip, the slates that form the roof courses and the hip are all in one plane.
- Cut the hip slates accurately to form tight joints and fill the joint with elastic cement.
- Place the nail holes so that they come under the succeeding hip slate.
- A variation of the mitred hip is known as the fantail hip.
- It is laid in the same manner as the mitred hip, but the bottom edges of the hip tiles are cut at an angle to form a fantail.

fastfacts

➤ *It is sometimes recommended that metal or slip flashings be woven in with each course at mitred hips.*

➤ *This is not usually necessary if proper care and workmanship are exercised when cutting, fitting, and embedding the hip slates.*

➤ *Some roofers do not use plastic or elastic cement on the hip slates and secure satisfactory results.*

Boston Hips

- Another popular type of hip is known as the Boston hip.
- The slates are woven in with the regular courses of the roofing slates, as shown in Figure 9.11.
- The nails then are covered with elastic or plastic cement and the lower part of the succeeding slate embedded therein.

Building Valleys

- Of the two methods used to form valleys, the first and most satisfactory is the open valley.
- The second, known as the closed valley, is considered by many to be more pleasing in appearance and often is used in high grade work.
- Variations of the closed valley, frequently used in connection with graduated or textured roofs, are the round valley and the canoe valley.

Open Valleys

- The open valley is formed by laying strips of sheet metal in the valley angle and lapping the slate over it on either side.
- This leaves a space between the slate edges that act as a water channel down the valley angle.
- The width of the valley, or the amount of space between the slate edges, should increase uniformly toward the bottom.

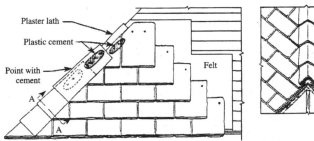

FIGURE 9.11   Boston hip.

- The amount of this increase, or taper, has been determined as 1 inch in 8 feet.

- For example, in a valley 16 feet long, the distance between slates should be 2 inches greater at the bottom than at the top because the width increases at the rate of ½ inch in 8 feet on each side of the valley.

- This permits a uniform flashing width that is approximately two-thirds the slate width under the slate adjacent to the valley.

- The difference in width allows the slate to be laid closer to the valley at the upper end than at the lower end and takes care of the increased water flow at the base of the roof.

- This tapering of the valley also has the very practical effect of allowing any ice that might form to free itself and slide down as it melts.

- Allow for this increase in valley width when placing the flashing strips.

- Valley flashings are generally laid in pieces up to 8 feet long.

- Theoretically, the best way to achieve the taper is to taper the sheets.

- Since this involves additional labor and material expenses, it often is more practical to use sheets that are shorter than 8 feet and increase the width of each sheet by an amount sufficient to take care of the taper. In this case, the increase in the width equals 1/8 inch per foot.

- The increase in the widths of succeeding sheets of various lengths is given in Table 9.5.

Example of an Open Valley

- ✔ A valley is 19 feet long.
- ✔ The sheet extends 5 inches under the slates and is fastened by cleats.
- ✔ Because a 4 inch minimum under the slate is necessary, two 8 foot and one 3 foot length can be used.
- ✔ Starting at the top:
- ✔ The first sheet would be 2 + 2 + 5 + 5 + ½ + ½ = 15 inches wide.
- ✔ If the 3 foot sheet is used at the top, the first 8 foot sheet would be 15⅜ or 15½ inches wide and the second one 16⅜ or 16½ wide.

TABLE 9.5 Taper Sizes

Length of sheets (in.)	Increase in width (in.)	Length of sheets (in.)	Increase in width (in.)
24	$^1/_4$	60	$^5/_8$
30	$^5/_{16}$	66	$^{11}/_{16}$
36	$^3/_8$	72	$^3/_4$
42	$^7/_{16}$	84	$^7/_8$
48	$^1/_2$	96	1
54	$^9/_{16}$		

EXAMPLE: A valley is 19 feet long. The sheet extends 5 inches under the slates and is fastened by cleats. What width of sheets can best be used? Because a 4-inch minimum under the slate is necessary, two 8-foot and one 3-foot length can be used. Starting at the top, the first sheet would be 2 + 2 + 5 + 5 + ½ + ½ = 15 inches wide. If the 3-foot sheet is used at the top, the first 8-foot sheet would be 15⅜ or 15½ inches wide and the second one 16⅜ or 16½ wide.

- Start the slate 2 inches up each side of the valley center at the top and taper away from the center at the rate of ½ inch for every 8 linear feet.
- The metal flashing should be of sufficient width to extend up and under the slate at least 4 inches (preferably 6 to 8 inches), and as far as possible without being punctured by the slating nails.
- When the two roofs that form the valley have considerable differences in slope, or when the differing roof sizes create a large variation in the volumes of water delivered into the valley, crimp the metal or make a standing seam.
- This breaks the force of the water from the steeper or longer slope and prevents it from being driven up under the slate on the opposite side.
- Condensation that forms on the underside of valley flashings, when not free to run off or evaporate, can attack the metal.
- It is recommended that the felt be omitted under the metal, unless copper is used.
- If felt is used under other metals, the metal should be well painted on the underside.

- For inexpensive roofs, the copper for valleys is laid flat, without crimps or cleats.
- For high grade work, secure the copper sheets to the roof boards and over the felt with metal cleats set 8 to 12 inches apart.
- Turn over the edge of the sheet ½ inch and hook it under the bent end of the cleat.
- Then nail the cleat to the roof boards with two nails and bend the cleat over to cover the nails.
- A high quality method, sometimes used on wide valleys, is to fold the metal 4 or 5 inches from the valley line and 3 inches from the cleat fold.
- This is known as foldover flashing.

Closed Valleys

- The closed valley is formed with the slate worked tight to the valley line and the pieces of metal placed under the slate.
- The size of the sheet is determined by the length of the slate and the slope of the adjoining roofs.
- Extend each sheet 2 inches above the top of the slate on which it rests so that it can be nailed along the upper edge of the roof sheathing without the nails penetrating the slate.
- Each sheet should be long enough to lap the sheet below by at least 3 inches.
- Set back the sheet from the butt of the slate above so that it is not visible.
- Separate these sheets by a course of slate.

fastfacts

Some roofers form the sheets with a center crimp, which stiffens them and forms a straight line on which to set the slates. This prevents water on one slope from forcing its way above the sheet and onto the other slope.

- Each sheet must be wide enough so that the vertical distance from the center of the valley to a line that connects the upper edges of the sheet is at least 4 inches.

- This dimension depends on the nailing of the slate, which should not penetrate the sheets.

- Another method of forming a closed valley is shown in Figure 9.12.

- The sheets are laid in long pieces directly on the paper or felt that covers the roof sheathing before the slate is laid.

- The sheets can be of any desired length and should lap in the direction of the flow by at least 4 inches.

- Nail the sheets about every 18 inches along the outer edge.

- Take care to avoid penetrating the sheet when nailing the slate.

Round Valleys

- The round valley forms a pleasing transition between two intersecting slopes when used in connection with the graduated or textured roof.

- The round valley requires a suitable foundation to establish the general contour.

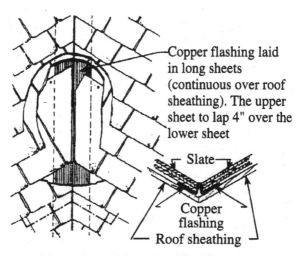

Copper flashing laid in long sheets (continuous over roof sheathing). The upper sheet to lap 4" over the lower sheet

Slate

Copper flashing

Roof sheathing

FIGURE 9.12 Forming a closed valley.

fastfacts

➤ *If not properly laid out, round valleys can produce disaster and mar an otherwise beautiful roof.*

➤ *For this reason, only experienced workers should lay this type of valley.*

➤ *It requires the most careful workmanship and a veteran craftsperson's knowledge of the situation to produce a job that is both pleasing in appearance and watertight.*

- The building of this foundation is usually considered a sheathing/deck function and might or might not be the job of the roofer.
- The construction and use of the foundation, however, is of great interest to the roofing contractor.
- There are three methods of constructing round valley foundations:
- First method:
  - Suitable for valleys of slight curvature, usually consists of a 12 inch wide board, with tapering sides, that is nailed into the angle formed by the intersecting roofs.
- Second method:
  - Suitable for any curvature or radius desired, 3 inch blocks are cut to fit the valley angle and sawn to the proper radius.
  - Then the blocks are nailed over the roof sheathing and spaced approximately the same as the exposure of the slates.
  - The blocks form nailing strips under the slate.
  - The size of the blocks varies due to the diminishing size of the valley as it approaches the ridge.
- Third method:
  - This is usually the most satisfactory method.
  - It is a combination of the first two.
  - The 3 inch nailing blocks are spaced from 20 to 30 inches apart over the regular sheathing.

fastfacts

All valley methods require a solid, accurately formed foundation that supports the slates and establishes the desired shape of the valley.

- Tapered strips ½ x 2 or 3 inches wide are nailed over the blocks for the length of the valley.
- For the installation itself, the valley slates must be at least 4 inches longer than the slate used in the corresponding roof courses.
- Trim the sides of the slates to the proper radius and shoulder the tops to make the slates lie flat.
- The round valley slates are sometimes embedded in elastic cement.
- If proper care is used when trimming and fitting the slate, no flashings should be necessary.
- Where the workmanship is not dependable, use flashings of metal or prepared roofing cut to the proper radius as a precautionary measure.
- But remember, when installing slate, the use of poorly trained mechanics should be avoided with diligence.
- Always use flashings wherever ice can form.
- The radius of the round valley is at maximum size at the eaves and gradually diminishes to practically zero at the ridge.
- For appearance, as well as to facilitate laying the valley slates, the distance across the eaves should not be less than 26 inches.
- If the roof condition does not permit this, use the canoe valley.

Canoe Valley

- The canoe valley is a variation of the round valley and is laid in the same manner, except that the radius at the eaves and ridge is practically zero.
- The radius gradually increases until it reaches maximum size half-way between the eaves and the ridge.

Eaves and Gables

- Start the slate under the eaves on a cant strip of a thickness suitable to the thickness of the slate.

- This enables the second course of slate to be laid correctly.

- In the case of a cornice, the slate should project about 2 inches beyond the cant strip, sheathing, or finishing member.

- The length of the slate under the eaves is determined by adding 3 inches to the exposure being used on the field courses of slates.

- Thus, if 16 inch slates are used, the exposure is 6½ inches and the size of the required slate is 9½ inches.

- Half slates are sometimes used, or roofing slates of the proper width can be laid horizontally.

- If the first course is ¾ inch thick, use ⅜ inch slates for the under-eave course or ¼ inch slates if the starters are ½ inch, although the under-eave and first course are sometimes made the same thickness.

- The first course of slate is laid over the under-eave course with the butts of both courses flush and the joints broken.

- When changing from a roof with a flat slope to one with a steeper slope, as in the case of a gambrel roof, project the slate of the upper and flatter roof 2 to 2½ inches beyond the steeper roof below.

- Use a cant strip to start the slate on the roof of lesser slope, the same as at the eaves.

- At the gables, the slate should overhang the finishing member of the verge board by not more than ½ inch.

- This dimension can be increased when close-clipped gables are used or the construction is such that the gable slates have ample nailing.

- The projection ought not to be too great for good appearance.

- Also, there are many interesting ways to lay gable-end or barge slates under regular courses along gable ends where shadow effect is desired.

- The ways and means of using and securing all gable-end slates depend on the type of construction.

Applying Flashing

As in all steep-slope applications, flashing must be used at all intersections of vertical or projecting surfaces or those against which the roof abuts, such as walls, parapets, dormers, sides of chimneys, etc. Flashings used over or under the roof covering and turned up on the vertical surface are known as base flashings. Metal built into the vertical surface and bent down over the base flashing is termed a cap flashing or counterflashing.

Base Flashing

- Extend base flashings under the uppermost row the full depth of the slate, or at least 4 inches over the slate immediately below the metal.
- Turn up the vertical leg not less than 4 inches and preferably 8 inches on the abutting surface.
- Where a vertical surface butts against the roof slope, build in the base flashing as each course of slate is laid. Turn out 4 inches on the slate and at least 8 inches above the roof.
- If the roof stops against a stuccoed wall, secure a wooden strip 4 inches wide, with a beveled top edge, to the wall.
- Then turn the base flashing out over the slate at least 4 inches and bend it up vertically at least 3 inches on the board.
- Except in unusual cases, it is satisfactory to turn the base flashing out 4 inches on the roof surface and up on the vertical surface from 6 to 8 inches for either sloping or flat roofs.
- Protect soil pipes and ventilators with base flashing.
- Vent pipes require base flashing. Use either special sleeves or one of the numerous patented flashing devices.

Counterflashing or Cap Flashing

- When the base flashing is not covered by vertical slate, siding, etc., use a cap flashing.
- Build this member into the masonry joints not less than 2 inches, extend it down over the base flashing 4 inches, and bend the edge back and up ½ inch.
- Reglets in stone or concrete are usually about 1 inch wide and 1 inch deep.
- Lay and form the flashing in the bottom of the cut and thoroughly caulk with molten lead on flat surfaces or lead wool on upright work.
- After caulking, fill the reglet to the surface with elastic or plastic cement.

- Use flashing hooks to secure stepped flashings.
- Make the vertical legs tight with roofers' cement colored the same as the masonry.
- On the best work, these flashings are soldered.

Chimneys

- As shown in Figure 9.13a, the chimney's base flashing is woven into the slate course and extends up under the counterflashing by at least 4 inches.
- The counterflashing for a chimney on the slope of a slate roof must lap at least 2 inches.
- Figure 9.13b illustrates the proper installation for a chimney on the ridge of a slate roof.
- The cap flashing laps the base by at least 2 inches.

Dormers

- The built-in base for a dormer window is shown in Figure 9.14.
- The flashings are woven into the slate course with each flashing sheet lapping the next lower one by 2 inches.
- Figure 9.15 details the flashing features of a recessed dormer window.

Saddles or Crickets

- Where a chimney or other vertical surface breaks through the roof at a right angle to the slope, build a saddle or cricket to throw the water away from the back of the vertical member.
- If the roof is structured of wood, use light rafter construction covered with sheathing boards, paper, and sheet metal.
- If it is a large area that is exposed to prominent view, slate it the same as the other roof areas.
- The size of the saddle is largely determined by the roof condition. It is usually sufficient to make the slope of the saddle the same as that of the roof.
- It is most important that the saddle or cricket be of adequate size, of ample slope, and well flashed.

a

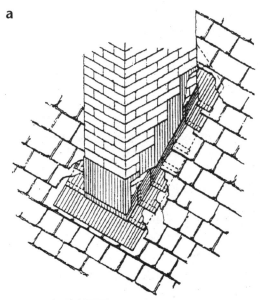

Built-in base flashing for chimney
on slope of slate roof

b

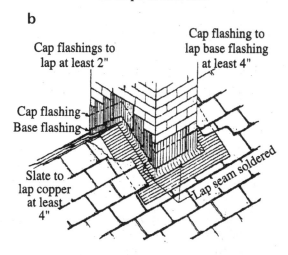

Cap flashing to
lap base flashing
at least 4"

Cap flashings to
lap at least 2"

Cap flashing
Base flashing

Lap seam soldered

Slate to
lap copper
at least
4"

FIGURE 9.13 (a) Chimney base flashing. (b) Proper
installation for chimney on ridge of a slate roof.

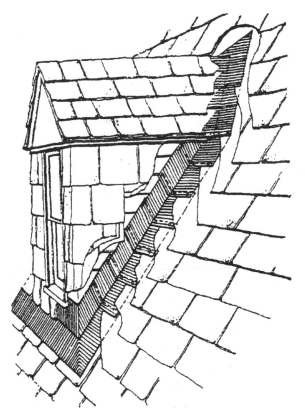

FIGURE 9.14 Built-in base for dormer window.

REROOFING WITH SLATE

Slate possesses qualities that make it suitable for reroofing any type of building. Its neutral or more vibrant colors quickly assume the characteristics of age, blend with the natural surroundings, and, with proper selection, conform to any desired color scheme. Its texture meets the demands of any design, and its cost is only slightly more than that of less permanent roofing materials.

• Removing the old covering exposes the sheathing or roof lath and permits a thorough inspection.

FIGURE 9.15 Flashing features of recessed dormer windows.

- Replace any broken boards or lath with new whole material and nail any loose boards into solid wood securely.

- Take out and replace original boards so that the joints are not broken over bearing rafters.

- As an alternative, cut in short rafters or blocking to act as bearings.

- Fill and secure any low spots or loose areas.

- When possible to do so, inspect the sheathing or lath from the underside and repair all broken or loosened boards.

- It is important to go over the roof and remove or drive home any projecting nails.

- Cut down any warped or raised edges or ends of sheathing or lath.

fastfacts

➤ *Sometimes the rafters of older buildings are of ample strength to support the present roof covering, but not designed to carry the additional snow or wind load recommended by present day engineering practice or required by the local building code.*

➤ *When the existing roof appears to sag or there are indications that existing supports are not adequate for the present roof covering, it is unwise to replace the old covering or to cover the roof with any new material before strengthening the roof supports.*

- Before laying the felt, thoroughly sweep off the sheathing to remove all chips, blocks, and loose nails.
- Rafters adequate for wooden shingle roofing are of sufficient strength for slate of commercial standard (³⁄₁₆ inch) thickness.
- When ordering nails, make allowances for the thickness of the old roofing and use a nail of sufficient length to secure penetration into the roof boarding.
- Note the width of the roof boards or lath spacing and order slate of proper length to secure a nailing that avoids the joints.
- Table 9.6 is useful for nailing.

MAKING ADDITIONS OR ALTERATIONS TO SLATE ROOFS

A properly laid slate roof requires little upkeep or care. However, houses are sometimes enlarged or remodeled. In such cases, it is often necessary to join new and old roofs or to remove and alter sections of the existing roof. There are several points to remember:

- It is desirable and necessary that the altered or additional roof matches the existing one in both shade and texture.

TABLE 9.6 Lath Spacing and Slate Length

Spacing of lath (O.C.) (in.)	Length of slate (in.)
$10\frac{1}{2}$	24
$9\frac{1}{2}$	22
$8\frac{1}{2}$	20
$7\frac{1}{2}$	18

- To obtain this result, attempt to secure slate of the same quality and color as the original slate.
- In many remodeling jobs, however, the slate has been on the roof for many years and no record is available of the quarry from which it came.
- In this case, the best way to proceed is to remove small adjoining sections and relay the slate, mixing some new slate with the old.
- This prevents a clear line of demarcation where the new work adjoins the old.
- In minor alterations, such as adding or removing a dormer, the old slate that was removed can be used again.
- In dormers and other projections, the lights and shadows differ from those on a roof expanse, which makes it easier to add new slate.
- For example, it would be best to use new slates on the new dormer cheeks and the old slates on other parts.
- Some roofers buy up a number of old roofs from buildings that are being torn down.
- They keep the old, weathered slate in their yards so that the slates can be readily matched to slates when minor alterations are made.
- Due to unavoidable causes, however, slates are sometimes broken on the roof.
- The broken slate can be repaired by the roofer.
- The best method is to first remove the broken slate.
- Cut the old slate with a ripper.
- Remove any remaining small pieces of slate (Figure 9.16).

fastfacts

➤ *Only responsible and experienced slate roofers should make alterations to a slate roof.*

➤ *If other workers are required to use ladders or scaffolds on slate roofs, use boards under the legs or uprights to distribute the pressure.*

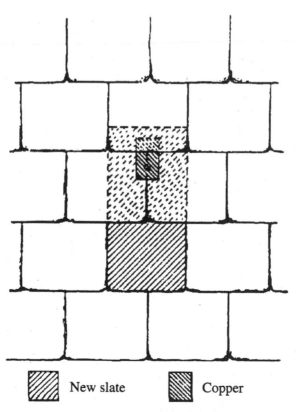

New slate Copper

FIGURE 9.16 Proper method of inserting a new slate.

- Insert a new tile and nail this slate through the vertical joint of the slates in the overlying course approximately 5 inches from the head of the slate, or 2 inches below the tail of the second course of the slate above.

- Over this nail, insert a piece of copper approximately 3 inches in width and 8 inches in length.

- Insert the piece of copper under the course above, lengthwise, so that it extends a couple of inches under the succeeding course.

- This ensures a proper lap and protects the exposed joint in which the nail is driven.

- Before installing, slightly bend this small piece of metal to ensure that it remains tightly in place.

OVERCOMING SLATE PROBLEMS

One of slate's major advantages is that it needs no ongoing maintenance—no painting or preservative coatings, and no cleaning. Slate resists seasonal weather changes better than other roofing materials, though some slates have a greater porosity than others and eventually begin to spall due to freezing cycles. Any roof, however, should be checked and maintained periodically.

Make maintenance of slate roofs a cash flow by signing a permanent contract with your customer. Chances are good that you will never put another roof on the building.

- Gutters and flashings are particularly prone to problems and may need occasional repairs.

- Complete failure of a slate roof is almost always due to poor installation methods, bad flashing details, or inferior nails.

- The nails sometimes give way.

- The worst condition is when all the nails need replacing because false economy or ignorance led to the use of the wrong nails.

- If some slates are letting go because their nails have rusted through, this could mean that eventually all the slates will have to be relaid with the proper copper nails.

- Galvanized nails are not recommended for slate roofs. When so much money is already being spent it, is wise use only waterproof nails.

- Remember that old slates can be recycled.

- Leaks in slate roofs are usually caused by deteriorated flashing or missing slates.

- Flashings gradually erode due to ice and other weather conditions.

- Replacing flashing in closed valleys, where the metal is covered by slates, must be done carefully, preferably by seasoned mechanics.

It is more economical to keep up the repairs on a sound slate roof by replacing missing slates and deteriorated flashings than to replace the roof or cover it with a modern, less permanent material. If the majority of the slates are delaminating or crumbling, it is not possible to save the roof. Explain to the customer that such a roof is probably many years old, and that the condition probably resulted from the original installation of inferior, less expensive slates.

Here, again, is an excellent opportunity to gain a maintenance business client. Owners with slate roofs typically care a great deal about their houses, welcome information, and can afford to maintain their homes.

If trees and vines are present, the leaves and branches can build up in valleys and dam water, which backs up under the slate. Maintaining drainage without breaking slates is an excellent service.

chapter 10

CLAY AND
CONCRETE TILES

C lay tile is still among the most popular roofing materials in
Europe, where homes and centers of commerce are designed
to last for generations. Their simplicity of form and shape
makes traditional, tapered mission tiles ideal for funneling and shedding water from pitched roofs. Concrete replicas have experienced
a dramatic surge in popularity in the last few decades.

Modern extrusion and pressed formed processes and high tech
gas fired kilns have replaced the primitive method of shaping clay
tiles over human thighs and then either baking them in the sun or
using wood fired beehive kilns. With these advances in manufacturing have come tremendous improvements in performance, quality,
and product diversity.

Manufacturing the Tiles

- The manufacture of clay tiles starts with shale that is crushed to a
  fine powdery clay.

- The clay is mixed with water and kneaded, or pugged, to the con-
  sistency of cookie dough.

- The clay is extruded through a die, like dough through a cookie
  press, and sliced into lengths.

- More complicated shapes come from pressing the clay into molds.

- Some ornamental tiles are even sculpted by hand.

- The formed tiles dry in a room kept at a temperature of approximately 90° to 95°F.
- From there, they go to a kiln for firing.
- For their trip through the kiln, the tiles are either stacked on a refractory, which is similar to a railroad car, and pulled through a tunnel kiln, or laid on ceramic rollers that convey individual tiles through a roller-hearth kiln.
- Regardless of the method, the object is to raise the temperature of the clay to the point of vitrification, which is about 2000°F.
- At this point, the clay minerals lose their individual identity and fuse together.

Coloring Clay Tiles

- High quality tile manufacturers achieve colors through the careful blending and mixing of various clays into complex clay bodies.
- Colors range from ivory and almond to deep reds and browns, apricot to peach and buff tones. Variegated accents are now available to designers.
- In addition, these colors can be enhanced by adding natural flash or variegated effects through the introduction of streams of natural gas during the firing process.
- By controlling the timing, frequency, and location of the kiln flashing, an infinite combination of randomly flashed tiles can be created for that truly custom, one of a kind color blend.
- Another method of coloring tiles is to spray a thin creamy layer of clay, called a slip, onto the tile before it is fired.
- The tile then takes on the color of the slip.
- The most dramatic, and most expensive, way to color tile is with a glaze.
- The metallic pigments in the glaze, when fired, melt to a glossy, vitreous, richly colored surface, much like that found on ceramic tile used indoors.

Benefits of Clay Tiles

- Premium clay roof tiles ensure protection from the elements and offer extended warranty periods.

- Some manufacturers offer lifetime warranties and even include fade coverage.
- Clay tiles offer the homeowner and roofing contractor numerous advantages:
  - Tile roofs typically last 50 years or longer and do not rust or otherwise deteriorate.
  - The color and texture of most tiles is integral and of natural materials that do not fade.
  - Along with durability, tile roofs offer the insulating quality of resisting the passage of heat gain from summer sun and winter heat loss.
  - The mass of tile roofing provides excellent insulation from sound infiltration.
  - Tile roofs are noncombustible and protect the structure from burning embers without suffering irreparable damage.
  - Tile and cement roof products carry Class A fire ratings.

Designing Tile Shapes

The roofing industry generally separates roof tile designs into three categories:

- High profile tiles:
  - The familiar mission, barrel, S, or Spanish influenced styles (Figure 10.1).
  - They are available in a variety of integral and applied colors.
  - According to legend, the curved shape of high profile tile evolved in ancient times when craftspeople formed wet clay sections over their knee to provide added stiffness.
- Low profile tiles:
  - They are manufactured in numerous different styles by several manufacturers (Figure 10.2).
  - They are also available in a variety of colors that complement any architectural style.
- Flat clay tiles:
  - These tiles have a shingle shape and are ribbed to simulate wooden shakes, or colored to represent slate.
  - Many of the flat clay tiles feature an interlocking system.

- Special clay tile shapes (Figure 10.3) are also available.
- For instance, closed-ridge end tiles and gable-terminal tiles are designed for gable roofs.
- Hip-terminal tiles are intended for decorative purposes and are used where a ridge and two or more hips intersect.
- Graduated tiles of diminishing widths are used for round towers, circular bays, and porches.
- Tile manufacturers furnish graduated tiles in many popular shapes.
- Some manufacturers also offer special valley tile, manufactured in angular or round form, and other special shapes for particular applications.

MAKING CONCRETE TILES

Concrete tiles are relatively new compared to clay tiles. Although concrete tiles have been used in Europe and Australia since the mid-1800s, they have only enjoyed widespread use in the United States since the mid-1960s. Concrete tile is composed of Portland cement, sand, and water, mixed in varying proportions. These materials are mixed and extruded on individual molds under high pressure to form the tile product.

Coloring Concrete Tiles

Concrete tiles are colored by one of two methods. The first is to add iron or synthetic oxide pigment to the batch mix. This produces a uniform color all the way through the tile. The second and less expensive method is to coat the tile with a slurry of cement and iron-oxide pigment. This technique also allows the manufacturer to add highlights of a second color, which creates a shaded or variegated effect.

- The particular coloring method used is dictated by several different requirements.
- The first of these relates to cosmetic and aesthetic appeal.
- The color coated product gives wider color hues, while the body colored product is less spectacular, is more subdued in appearance, and has a limited color range.

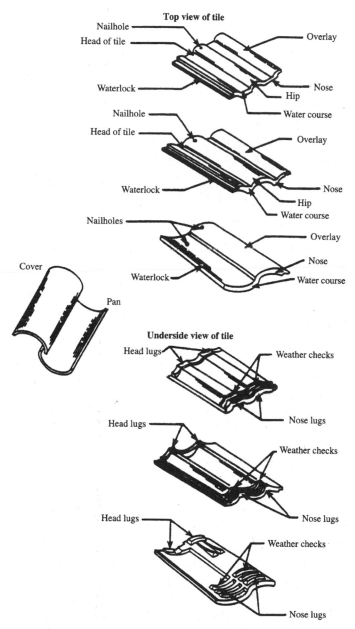

FIGURE 10.1 High-profile S tiles.

FIGURE 10.2 Low-profile tile.

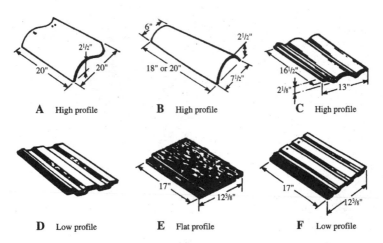

FIGURE 10.3 Special clay tile shapes.

- Application requirements for both product types relate primarily to atmospheric and climatic conditions, as well as aesthetic and architectural intent.

- Experience has shown that the surface coated product is, generally speaking, more resistant to growth of, and discoloration from, moss and lichen found in areas with high humidity and in areas with large amounts of rainfall.

- The color coated product does not fare well in areas with extreme freeze and thaw conditions or where there is a large amount of industrial pollution that contributes to such phenomena as acid rain and deposits of atmospheric dirt and grime.

- The through color product is more resistant to freeze and thaw conditions and more subject to discoloration and staining by moss, lichens, and atmospheric pollution.

- With either coloring method, the tiles are usually sprayed with a clear acrylic sealer.

- The sealer helps the tiles cure properly.

- It also controls any efflorescence, which is the white powder consisting of free lime that surfaces as concrete ages.

- The acrylic sealer forces the lime out the underside of the tile, where it doesn't spoil the appearance.

- As a side effect, the sealer gives the tile a slight gloss.

- The gloss of the sealants of concrete tiles wears off in a few years, and the color softens to its true matte finish.

Designing Tile Shapes

Concrete tile has three classifications: flat, roll, and graduated. Flat tiles vary in size and have the appearance of slate or wooden shakes. Some of these tiles are made of fiber reinforced cement and come in various colors. They can be installed on roofs up to 40 feet high in areas with wind speeds up to 80 miles per hour (mph). Some flat tiles are available with interlocking water locks.

Roll tile is pan and cover shaped. It is better known as barrel or mission tile. Graduated tiles of diminishing widths are used for round towers, circular bays, and porches. Some tile manufacturers furnish graduated tiles in all popular shapes.

PREPARING THE ROOF FOR CLAY AND CONCRETE TILE

A new roof must receive the same preparation whether clay or concrete tiles are to be applied.

Matching Roof Slope and Underlayment

- Roll or flat tile can be applied to roof decks with slopes of 4 inches per foot or more when a minimum of one layer of 30 or 43 pound felt is applied horizontally to serve as the underlayment, and the tiles are nailed or wired with a minimum 3 inch headlap.

- Any style of clay or concrete tile can be applied on solid sheathed roof decks with slopes less than 4 inches per foot when a minimum of two layers of 30 or 40 pound nonperforated, asphalt-saturated felt are set in hot asphalt or mastic to serve as the underlayment.

- One layer of a modified-bitumen-coated sheet, with laps either torched or heat welded, is also acceptable.

- Install vertical lath stringers with horizontal battens fastened over the stringers over the underlayment.

- This creates a nailing surface over which the tile can be installed.

- The tile must be installed with a minimum 4 inch headlap.

Preparing Deck Surfaces

- If plywood is used as the deck material, use exterior plywood thick enough to satisfy nailing requirements.

- Separate the plywood panels by at least $\frac{1}{16}$ inch to allow for expansion.

- If wooden boards are used for the roof deck, the boards should be a minimum of 1 x 6 inches.

- They should span a maximum of 24 inches between trusses or rafters.

- When the roof deck is made of concrete, a surface must be provided onto which the tiles can be applied.

- To create this surface, run 1 x 2 inch, beveled wooden nailing strips of treated lumber from the eaves to the ridge.

- Embed the strips in the concrete and space them 16 or 24 inches on center.

- Nail the felt to these nailing strips.

- Then nail lengths of lath, applied vertically, directly into the beveled nailing strips through the felt.

- Finally, nail 1 x 2 inch battens, or stringers, spaced according to the type of tile to be used, horizontally across the lath.

- This simulates a wooden surface that can accommodate the application of tiles.

- Some concrete tile manufacturers also produce special wiring systems for securing concrete tile to concrete roof decks.

WORKING WITH TILES

When working with tiles, it's important to keep the following precautions in mind.

- Broken tiles:
  - More tiles are broken in transit and on the ground than are broken on the roof.
  - Take great care when unloading tile at the jobsite.
  - Unload the tile as near to the building as possible, and distribute it so that delivery to the roof is convenient.
  - Save tiles that have been broken either in transit to the roof or during application.
  - Use these tiles when cut tiles are required.
- Loading tiles:
  - To prevent tiles from breaking or becoming soiled, stack them not more than six high.
  - Keep nails, cement, and coloring material covered until needed.
- Cutting tiles:
  - When tiles must be cut, mark the desired break line on the tile.
  - Then carefully cut along this line.
  - If too much tile is cut at once, the tile can fracture.
  - The correct cutting procedure requires that the tile be placed on a tile stake well back from the cutting line and tapped with a hammer.
  - Further trimming should be done with a large pair of pincers.
  - Tile saws or power saws also can be used to cut tiles.
  - Diamond-tipped tile saw blades are best for this process.
- Narrow tiles:
  - Concrete tiles less than three quarters of the width of a full tile are susceptible to wind damage when used on gable ends.
  - Drill an additional nail hole in the top of the tile and place a dab of roofer's mastic under the butt end.
  - Mechanically fasten the tile through the newly drilled hole.
  - Gable-end, partial tiles installed in this manner can withstand the same windy conditions as the field tile.
- Drains:
  - To avoid choking drains with broken tile or tile trimmings, do not permanently install drains until tiling work is complete.
  - Keep drain outlets covered when tiling to prevent any debris from clogging the leader pipes.

Making Allowances for Cold Weather

• Freeze and thaw conditions are encountered in many areas of the United States.

• A freeze and thaw area is defined as one that experiences 30 cycles of freezing and thawing per year.

• One cycle encompasses a change in temperature from more than 32°F to less than 30°F that is accompanied by moisture that freezes into ice and thaws to a liquid state.

• Freeze and thaw failure can occur in just about every product, manufactured or natural, that is exposed to such conditions.

• Dry, cold weather temperature fluctuations above and below freezing have little or no effect on concrete tile.

• To protect tile from freeze/thaw conditions, apply a minimum of one layer of 40 pound coated felt horizontally with a minimum 4 inch headlap and 6 inch sidelap.

• Slopes below 4 inches in 12 require a functional builtup roof (BUR), a modified bitumen roof (MBR), or a self-adhering, ice-and-water-shielding bitumen membrane underlayment system.

• In addition to the 40 pound underlayment, the following is required as an ice shield, regardless of slope, on eaves and barges, or rakes.

• Starting from the eave and barge to a point 36 inches beyond the inside wall line of the structure, use one layer of 40-pound coated felt set in roofer's mastic or cold-process adhesive, or one layer of self adhering, ice-and-water-shielding bitumen membrane.

• Over the underlayments or decks, fasten a vertical counter-batten, at least 1 x 2 inches, at a minimum of 24 inches on center from the eaves to the ridge.

fastfacts

➤ *The counter-batten system minimizes condensation by allowing air circulation and, with proper ventilation, helps prevent ice-dam buildups.*

➤ *A minimum 4 inch tile headlap is recommended in areas with heavy snowfalls.*

fastfacts

Discuss quality control of material shading and uniformity of tile with the manufacturer, architect, and building owner prior to the contract and prior to placing an order for the product.

- Over the vertical counter-batten, fasten a second horizontally installed batten, at least 1 x 2 inches, spaced to ensure a minimum 3 inch tile headlap.
- Use only treated lumber.

Arranging Colored Tiles

Blending modern colored clay and concrete tiles can be a rather difficult procedure. To guard against a spotted colored roof, mix tiles in the correct color arrangement on the ground and then send them up to the roof in bundles along with strict application instructions.

- For example, if the color scheme calls for 10 percent of one color, 30 percent of another color, and 60 percent of a third color, send the tiles up to the roof in bundles of 10 tiles, with each bundle having one tile of the first color, three of the second color, and six of the third color.
- In this way, the tiles can be applied in the order in which they were bundled, and no time is wasted selecting colors on the roof.
- Separate the 10 tile bundles into two stacks of five tiles each when loading them onto the roof deck.
- After 75 to 100 tiles have been installed, visually inspect the applied tiles from ground level and at a distance from the building to ensure that the tile courses follow straight and true lines and that the colors of the tile blend well.
- Repeat this procedure at regular intervals during installation to ensure an attractive and acceptable roof.
- The blending of tile shades to avoid streaks or hot spots is particularly important.
- Pre-blended tiles can be obtained from some manufacturers.

Fastening Tiles

- Use nails and screws to attach clay and concrete tiles to decks.
- Use ⁵⁄₁₆ inch compression spikes for concrete decks and No. 12 TEK screws for steel decks.
- On a plywood deck, use ring-shank nails of sufficient length for slight penetration through the underside of the deck.
- For board plank decks, use smooth-shank nails at least 1½ inches long that do not penetrate the underside of the deck.
- For gypsum plank and nailable concrete decks, use stainless steel or silicon-bronze screw-shank nails of a length sufficient to penetrate ½ to ¾ of their length into the deck.
- Do not penetrate the underside of the deck. If the deck is excessively hard, use smooth-shank nails.
- Fasteners also can be used to hold tiles.
- Do not drive home fasteners or draw the tile.
- Drive fasteners to a point where the fastener head just clears the tile, so that the tile hangs on the fastener.
- When tiles are fastened too tightly, they lift up at the butt.
- This allows high winds to blow them off the roof or to blow water under them.
- On exposed overhangs, the fasteners should not penetrate the sheathing.
- Drive all fasteners into the roof sheathing and not between sheathing joints.
- This is especially important near the top and sides of the roof.
- When battens are used, drive all fasteners into the batten boards.
- Fasten tiles individually.
- Secure hip and ridge tiles with one nail in each tile and with a golf-ball-size dab of roofer's mastic under the tile at the headlap, recessed so that it does not show.
- Barge or verge tiles, when available, require two nails and roofer's mastic.
- Table 10.1 gives the nailing procedures for attaching clay and concrete tiles to plywood and wooden sheathing.
- Battens should consist of a nominal 1 x 2 inch approved material spaced parallel with the eaves to achieve a minimum 3 inch tile headlap.

- Use battens over solid sheathing and an approved underlayment.
- Make provisions for drainage at a maximum of every 4 feet past or beneath the battens.
- Battens must be attached with four corrosion resistant fasteners per batten.
- Over concrete decks, use wire tying strips instead of battens to secure the tiles.
- Secure angular strips of wire, 1½ x ½ inches in size, to the concrete deck with expansion bolts.
- Perforate the wide flange of the wire strip with holes spaced at regular intervals suitable to the tile exposure.
- Then run 14 gauge tie wire through these holes and tie it around the holes in the tile covers.
- Turn up the wire and twist it under the lap of the succeeding tile.
- When tiles are attached directly to metal purlins, fasten them with 14 gauge rust-resistant wire or self-tapping screws with a minimum ⁷⁄₁₆ inch diameter head.

TABLE 10.1 Attachment of Tiles to Sheathing

| | Field tile nailing | | Nailing for perimeter tile and tile on cantilevered areas[2] |
Roof slope	Solid sheathing with battens	Solid sheathing without battens[1]	
3/12 to and including 5/12	Not required	Every tile	Every tile
Above 5/12 to less than 12/12	Every tile every other row	Every tile	Every tile
12/12 and over	Every tile	Every tile	Every tile

1. Battens are required for slopes exceeding 7/12.
2. Perimeter nailing areas including three tile courses but not less than 36 inches from either side of hips or ridges and edges of eaves and gable rakes. In special wind areas, as designated by the building official, additional fastenings might be required.

- If self-tapping screws are used, they should be capable of penetrating a minimum of ¾ inch into the purlins.
- Either nail or wire each ridge and hip tile in place.
- If tiles are to be wired in place, lace 14 gauge rust-resistant wire through the nail holes in the tiles and securely tie it to the heads of nails driven into the ridge or hip boards (Figure 10.4).
- A golf ball size dab of roofer's mastic is required at the tile headlap.
- Recess it so that it does not show.
- Nails for tile roofs should be made of 11 gauge, rust resistant, aluminum, copper, yellow metal, galvanized, or stainless steel and be of sufficient length to penetrate either ¾ inch into the sheathing or through the thickness of the sheathing, whichever is less.
- On exposed overhangs, nails should not penetrate the sheathing.
- Where building officials have designated their localities high wind hazard areas, special fasteners must be used.
- In these wind hazard areas, secure the nose end of all eaves-course tiles with hurricane clips (Figure 10.5).
- Hurricane clips are available in different shapes to suit the type of roof sheathing used.

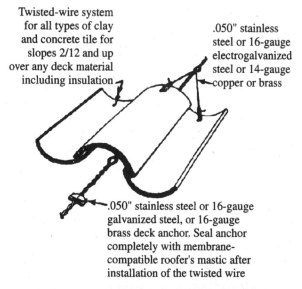

Twisted-wire system for all types of clay and concrete tile for slopes 2/12 and up over any deck material including insulation

.050" stainless steel or 16-gauge electrogalvanized steel or 14-gauge copper or brass

.050" stainless steel or 16-gauge galvanized steel, or 16-gauge brass deck anchor. Seal anchor completely with membrane-compatible roofer's mastic after installation of the twisted wire

FIGURE 10.4 No. 14 gauge rust-resistant wire.

fastfacts

➤ *On all roofs, over all roof slopes, and under all conditions, securely fasten all tiles installed on cantilevered sections of the roof, such as gables or eaves, and all tiles installed at the perimeter of the roof.*

➤ *Where tiles overlap sheet metal, secure them with appropriate tie-wire systems.*

- Lay the tiles with a minimum 3 inch headlap.
- Nail each tile to the roof sheathing with one 11 gauge, rust resistant nail with a minimum $\frac{5}{16}$ inch diameter head.
- Apply a bead of roofer's mastic over the nail heads that fasten gable, barge, and ridge tiles.
- On extremely steep or vertical roofs, wind currents can cause tiles to rattle.
- The recommended method for preventing rattling is to use hurricane clips.
- Another way to prevent tile rattling is to set the butt edge of each tile in a dab of roofer's mastic.

fastfacts

➤ *When specifications require that all tiles be embedded in plastic cement, cover all lateral laps with cement.*

➤ *Use approximately 40 pounds of cement per square.*

➤ *Continuously embedding tiles in plastic cement throughout the roof restricts roof movement. Cracked tiles can result because of the expansion and contraction of the roofing during changes in temperature. This method of application is not recommended.*

Use one per
tile on sidelap

FIGURE 10.5 Hurricane clips.

Applying Flashing

Basic tile flashing is applied in much the same manner as for slate and wooden shake roofs, which is fully described in Chapters 8 and 9.

- For valley flashing, use at least 28 gauge, corrosion-resistant metal.
- Extend it at least 11 inches from the centerline of the valley each way.
- Form a splash diverter rib, as part of the flashing, not less than 1 inch high at the flow line.
- Overlap flashing at least 4 inches.
- Valley metal and flashing should be in place prior to tile application.
- For other flashing, use at least 26 gauge, corrosion resistant metal.
- At the sides of dormers, chimneys, and other walls, extend the flashing at least 6 inches up the vertical surface.
- Thoroughly counter flash and extend the flashing under the tile at least 4 inches.
- Turn the edge up 1½ inches.
- Long runs of flashing material at parapet walls (Figure 10.6) and copings, where roof tiles come to an abrupt termination, can be made of rigid materials, such as 26 gauge galvanized sheet metal.
- If rigid materials are used, form them in such a way that they provide sufficient coverage and adequate drainage.

- Establish an acceptable wind block at longitudinal edges of flashings by grouting the longitudinal edges with Portland cement mortar or by using alternative materials acceptable to local building officials.

- Flashing around roof penetrations should be in place prior to the application of tiles.

- Flashing metal should be at least 28 gauge galvanized metal or an equivalent noncorrosive, nonstaining material.

INSTALLING THE TILE ROOF

Examine the areas and conditions under which the tile is to be installed with care. Do not proceed until unsatisfactory conditions have been corrected. To avoid later disputes, report any such conditions to the contractor and other subcontractors in writing, and keep these letters on file.

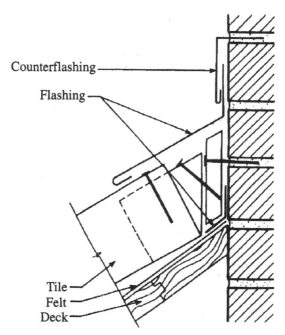

FIGURE 10.6 Flashing material at parapet walls.

Before Starting an Installation

- Verify that deck surfaces are clean and dry.
- Remove all foreign particles from the substrate to assure proper seating and to prevent water damage.
- Make certain that your installation will incorporate the specified tiles in strict accordance with pertinent local code requirements.

The Installation

- On vertical applications, and on extremely steep pitches where wind currents can cause lift, set the butt of each tile in a bead of the specified plastic cement or sealant, or provide copper hurricane clips at intervals.
- Use plastic cement and sealant with care, and avoid smearing the exposed tile surface.
- Chalk horizontal and vertical guidelines on the membrane as a pattern to assure water tightness and proper appearance.
- Space the chalklines by measuring the delivered tiles for average length and width exposures.
- Do not exceed an average exposure length of ¼ inch.

Applying Flat-Tiles

- Mark off the roof horizontally.
- Vertical lines, marked off randomly, help maintain a good vertical alignment.
- For roofs with pitches of 4/12 and above, install 9½ inch eave blocking (which is available from the manufacturer), eave metal, a bead of sealant, and 40 pound felt.
- Lay the felt parallel to the eave metal and extend it ¼ inch over the lower edge (Figure 10.7).
- Note that some local building codes require an ice and water shield along the eaves.
- For roofs with pitches below 4/12, install 12 inch eave blocking, eave metal, and a minimum 3 foot wide strip of ice and water shield along all the eaves.
- Lap the felt 5 inches instead of the standard 3¾ inches and install battens that are notched 8 inches on center (Figure 10.8).

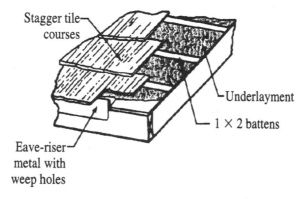

Stagger tile courses

Underlayment

1 × 2 battens

Eave-riser metal with weep holes

FIGURE 10.7 Flat-tile application.

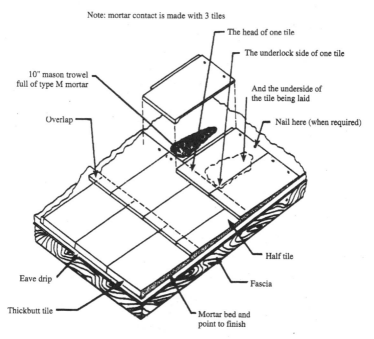

Note: mortar contact is made with 3 tiles

The head of one tile

The underlock side of one tile

10" mason trowel full of type M mortar

And the underside of the tile being laid

Overlap

Nail here (when required)

Eave drip

Half tile

Fascia

Thickbutt tile

Mortar bed and point to finish

FIGURE 10.8 Flat-tile application, pitches below 4/12.

fastfacts

On reroofing jobs, or when the eave fascia is not raised, install eave blocking with a cant strip and use new eave metal.

- When laying out the roof, install a ridge nailer (vent).
- Next, strike lines that are centered on each hip, if applicable.
- Then strike the horizontal line for the top edge of the first batten 13¼ inches above the eave for a typical 15⅜ inch tile to ensure proper fit of the bottom row of tile.
- Next, strike the horizontal line for the top edge of the last batten so that the field tile butts the ridge nailer (vent), which is approximately 1 inch below.
- Then divide the distance between these two lines into equal increments not to exceed 12 inches and strike lines for the top edges of the battens.
- If different eave lines do not allow for equal spacing, overlap the bottom row to allow the second row to match the rest of the equally spaced rows.
- If a short row is required, lower the height of the eave fascia board and use ½ x 16 inch battens with 1 inch spacing for the bottom row.
- Doing so allows less than a 10 inch exposure on the bottom tile without causing the second row to lie at a different pitch from the rest of the tile.
- If different ridge heights do not allow for equal spacing, add a short row along the shortest ridge.
- If a short row is required, cut off the head of the tile, drill a new nail hole, and install a thicker batten tight to the ridge nailer so that the short row maintains the same pitch as the rest of the tile.
- Install hip nailers to within 6 inches of the bottom corner and then install the horizontal, 1 x 2 inch batten strips, leaving 1 inch spaces between the ends.
- Use pressure treated 1 x 2 x 8 inch battens with notches or ports 16 inches on center.
- On pitches 4/12 and above, install with 18 fasteners.

- On 3/12 to 4/12 pitches and vented cold-roof applications, use pressure treated 1 x 2 inch x 6 foot battens with notches or ports every 8 inches.

- Use noncorrosive fasteners of sufficient length to fully penetrate the roof sheathing.

- Another approved procedure is to use 1 x 2 inch horizontal batten strips, without notches, installed over 1 x 2 inch pressure treated vertical battens 16 inches on center or 1 x 4 inch horizontal battens installed over 1 x 2 inch, pressure treated vertical battens 24 inches on center.

- For flat tile on gable roofs, put an X on the third batten above the eave, 35 inches on center, starting at the left gable edge.

- Randomly place stacks of four tiles above each X on every other batten.

- Then stack four more tiles randomly on top of the existing stacks to get a good color blend across the entire roof position.

- Put one barge on each batten next to the gable edge and install with gable tile, or later from a ladder or staging after the field tiles are in place.

- To lay flat tile on hip roofs, place the Xs at 35 inches on center on the third batten above the eave and stack the tile above each X.

- Add extra stacks along the hip and valley to allow for 1½ tiles per row to be cut on each side.

- Install ridge trim as the roof progresses.

- Lay the trim with approximately a 17 inch exposure and seal between the laps with mortar (Figure 10.9).

---

fastfacts

➤ *When using vertical battens, remember to raise the eave and fascia metal to the additional thickness of the vertical batten.*

➤ *Fill all hip, ridge, and other voids with American Society for Testing and Materials (ASTM) C-270 Type M cement mortar and neatly point it.*

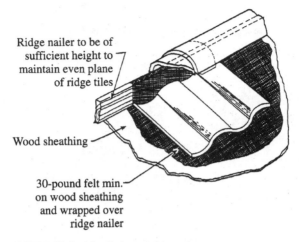

Ridge nailer to be of sufficient height to maintain even plane of ridge tiles

Wood sheathing

30-pound felt min. on wood sheathing and wrapped over ridge nailer

FIGURE 10.9 Installation of ridge trim.

Applying Roll Tiles

- When roll or mission tile is applied, an under-eaves course of tile does not need to be applied.

- Apply the first course of roll tile above an eaves strip.

- This metal eave-closure strip, better known as a birdstop, is specially formed to fit the underside of the tile.

- They are fastened 1¼ inches back and inside the cover of the tile.

- Apply the balance of the tile with the head of the tile aligned with the horizontal guidelines.

- Tile spacing may need to be adjusted to provide a uniform exposure.

- Load the tiles on the roof so that they are evenly distributed and within easy reach.

- Extend the first row of tiles ½ to ¼ inches over the eaves.

- For example, when a 13¾ inch tile is used, strike the first horizontal guideline 12 to 12¾ inches from the eaves line.

- Where chimneys or dormers project through the roof, loosely lay the first course of tile along the eaves.

- To minimize tile cutting, secure the tiles to the eaves only after making adjustments for the projections.

fastfacts

➤ *The procedure for installing mission tile varies according to the tile design and size.*

➤ *Each manufacturer covers the specifics in its installation instructions.*

➤ *Basically, as with clay and concrete flat tiles, the roof is first felted and then horizontal and vertical guidelines are chalked to indicate the courses.*

➤ *Layout is critical because any deviations stand out against the pronounced vertical pattern.*

➤ *If the tile is designed with lugs to hang on battens, nail those battens next.*

➤ *Some manufacturers approve of hanging their tiles on spaced sheathing over heavy felt underlayment that is draped over the rafters.*

• Cut or weave tiles as they are installed down each side of a valley.

• Valleys can be open, mitred, swept, rounded, or closed with special tile.

• Special valley tile produces much the same effect as the rounded valley tile.

• It is longer than regular tile and is fan-shaped.

• Because of its shape, valley tile need not be nailed, but should be cemented at the laps.

• When valley tiles are used, first extend a row of tiles up the valley.

• Tiling should then proceed back toward the valley from the verge or gable.

• The last tile should be large and trimmed against the valley tile.

• Fit tiles that converge along the hips of a roof close against the hip board.

• Make a joint by cementing the hip tile to the hip board with roofing cement or mortar.

• Color the cement to match the tile.

- Notch these cut tiles and either nail or wire them to the hip board.
- It is advisable to lay a golf ball size dab of roofer's mastic between the tiles.
- Begin the hip roll with a hip starter, which is a hip roll with one end closed, or a hip stack, which is a stack of hip roll pieces equal in height to the hip stringer.
- Nail the hip-starter tile to the hip board with nails of appropriate length and follow with the regular hip roll, lapped either 3 inches or in accordance with the manufacturer's requirements.
- Cement between the laps.
- Do not fill the interior spaces of hip or ridge rolls with pointing material, as this material inhibits air circulation.
- Cover ridges in much the same manner as hips.
- Fill the spaces between the tiles in the top row with special ridge fittings or with cement mortar colored to match the tile.
- When tile fixtures are used at the ridge, nail the diagonal half of a 2 x 4 on either side of the ridge board to provide a nailing surface for the tile fixtures.
- This is not required when Portland cement mortar is used as fill material between the last course of tile at the ridge.
- Some tile manufacturers make batten strips available for these layouts, while others provide detailed drawings with which contractors can make their own battens.
- Spanish or S tile designs:
  - Mark off the roof vertically and horizontally.
  - Interlocking, not-lugged tile can be laid with a minimum 2½ inch headlap.
  - Lugged tile should maintain a 3 inch design for mortar application.
  - Check with the manufacturer of the particular tile.
- Prefabricated birdstops or eave closures:
  - Prefabricated eave closure strips or mortar can be used to elevate the butt end of the first, or eave, tile to attain the proper slope.
  - When using mortar, provide weep holes next to the deck to allow proper drainage of any moisture accumulation under the tiles.
  - Place a full 10 inch mason's trowel of mortar under the pan section of each tile, beginning at the head of the tile in the preceding course.

- Press each tile into the interlocking position so that the cover rests firmly against the lock of the adjacent tile.
- Installation details for eaves, ridges, gables, and so forth are given in Figure 10.10.
- Two-piece roll or barrel mission layouts:
  - Mark off the roof vertically and horizontally.
  - Maintain a minimum 3½ inch headlap.
  - Use mortar or prefabricated eave-closure strips or birdstops to elevate the butt end of the first, or eave, tile to attain the proper slope.
  - Provide weep holes next to the deck to allow proper drainage of any moisture accumulation under the tiles.
  - For roof pitches of 3/12 and steeper, provide a minimum of one layer of 30 pound felt or upgraded material.
  - Install a birdstop or a 1 x 2 inch wooden strip to boost the first course of tile.
  - Then install the first row of tile leaving a 3 inch overhang.
  - Use one corrosion resistant nail not less than a No. 11 gauge, $\frac{5}{16}$ inch head per tile or use the wire tie system.
  - Then lay a booster tile above the birdstop or cave strip, followed by the starter tiles. Other details necessary for laying a barrel/mission roof are given in Figure 10.11.
- Oriental Style:
  - Figure 10.12 shows a typical Oriental roll-style tile layout and installation details. Without the use of ornaments, this tile is used in Western contemporary designs.
  - While the most traditional color for a Japanese-style teahouse or temple is black, these Oriental style tiles are available in natural red and glazed colors.

Turret Tile

True turret roof designs or fan-shaped applications are now possible without compromising design concepts. To determine the quantity of tile needed and specific installation guidelines for a given job, ask the manufacturer to provide a scale drawing or blueprint of the top and side views. All that the manufacturer needs is the diameter of the circle and roof pitch (Figure 10.13).

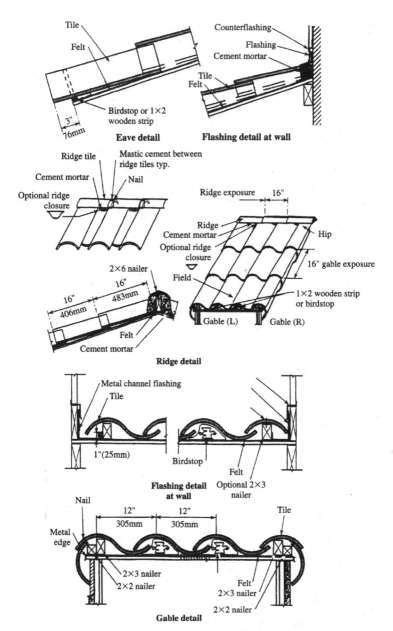

FIGURE 10.10 Installation details for eaves, ridges, and gables.

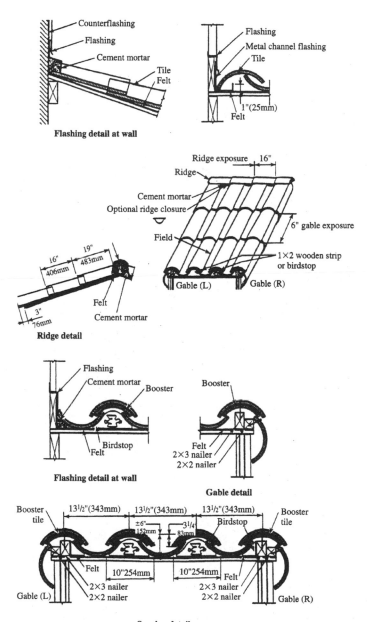

FIGURE 10.11 Laying barrel/mission tile on roof.

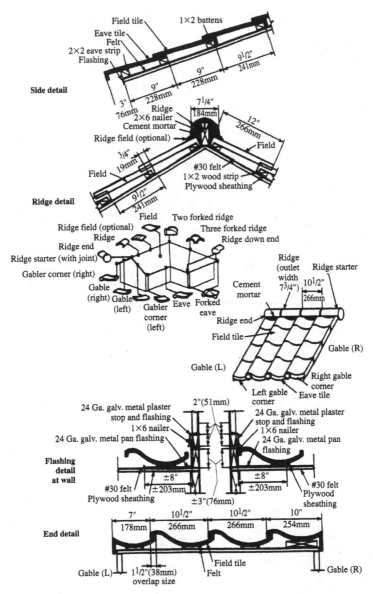

FIGURE 10.12 Typical oriental roll style tile layout and installation details.

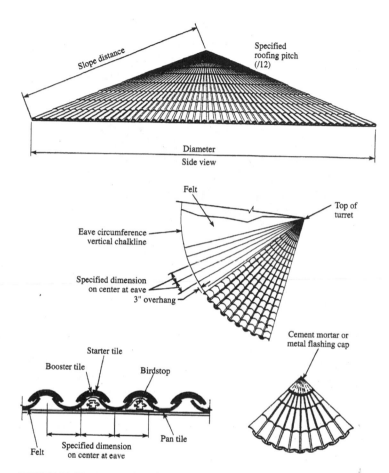

FIGURE 10.13 Turret tile—diameter of the circle and roof pitch.

- General instructions for installing turret tiles:

  - At the first course, between vertical chalklines, install a clay bird-stop and then place the pan tile on top of the vertical chalkline.

  - Fasten each pan tile with a copper or other noncorrosive 11 gauge, large-headed nail, or use the wire tie system.

  - If the job site is located in a high wind area, use mortar or other sealant to secure the pan tile.

- Once the birdstop and pan tiles are in place, install the booster and starter tile.
- Secure with copper wire or noncorrosive nails.
- For the rest of the courses, lay 16 inches to the weather.
- When the tile becomes crowded, adjust to the next smaller size and continue to the top of the roof.
- Follow the chalkline and use the turret worksheet provided by the manufacturer.
- Prior to installing the last two or three courses, lay a mockup to assure proper fit.
- Do not use adhesives or nails to secure the tiles until the mockup is complete and satisfactory.
- Note that the final two or three courses normally lose one to two lines, or more, close to the top.
- Start installing the final two or three courses from the top down and secure each tile.

INSTALLING LOW-SLOPE, MORTAR-SET ROOFS

Roofing tiles have been installed in mortar for centuries. The practice of installing tiles with mortar over a built-up subroof evolved in high wind and high moisture areas of the southeastern United States.

In this system, the built-up subroof provides the moisture barrier and the tiles, in addition to being aesthetically pleasing, protect the subroof from the sun's ultraviolet (UV) rays, high winds, and external damage. The system also allows the use of tile on lower sloped roofs.

fastfacts

➤ *After any style clay or concrete tile roof is laid completely, do not allow traffic on the roof that might vibrate the framing or roof sheeting.*

➤ *At least 24 hours are needed to ensure a proper set. Prohibit roof traffic for at least 72 hours.*

- Apply the mortar over a solid sheathing of at least ⅝ inch plywood or 1 inch tongue and groove.

- Mechanically fasten one layer of at least 30 pound organic felt underlayment to the sheathing.

- After applying the first layer, install metal eave flashings.

- Next, apply one layer of mineral surfaced rolled roofing material to the underlayment with hot asphalt or mastic and then backnail.

- Other mineral surfaced products can be used, such as MBR, although care should be taken to guard against roof slippage.

- Mortared tile can be used on slopes down to a minimum of 2 inches in 12.

- On slopes between 5 inches in 12, and 7 inches in 12, additional mechanical fastening is required for the first three courses in areas subject to high winds.

- On slopes 7 inches in 12 and steeper, mechanically fasten all tiles.

- Mortar used to adhere tile to the sunroof should be as specified in ASTM Specification C-270 Type M.

- Soaking the tile prior to installation and/or adding additives to the mortar might be required to achieve proper adhesion between mortar and tile.

- In areas of the country subject to blowing sand or heavy rainfall, use mortar at ridge or hip intersections to provide a weather block.

- Use mortar sparingly and only to provide proper bedding for hip or ridge tiles. Specially designed metal weather blocks are available from most manufacturers.

REROOFING TILE ROOFS

Roofing contractors, architects, and specifiers, particularly those in the sunbelt or western states, sometimes encounter a reroofing project that involves a clay tile roof that is 70 years old or more.

Their first impulse might be to draw up specifications for the job on the assumption that the existing tile needs to be entirely removed and replaced with new material.

This thinking is understandable. UV radiation, heat, moisture, and exposure to the elements work together to limit the life span of most roofing materials to 25 years or less.

fastfacts

➤ *First quality clay roofing tiles last indefinitely if the roof is properly laid and maintained.*

➤ *There are many old roofs that remain basically sound, with the clay tile intact and the underlayment in generally good repair.*

➤ *When these roofs begin to leak, it is often a result of problems with the underlayment.*

➤ *The tile often can be used again.*

➤ *Even if the existing tile is in good condition overall, there can be broken or damaged tiles on the roof.*

Failure to understand the long lasting nature of the best clay tiles could be a costly mistake. A simple computation based on the money saved in new materials, plus the life cycle cost benefits offered by tile roofing, normally results in a reroofing specification based on lifting and relaying the existing tile.

- The first step in assessing the condition of the roof:
  - This can be conducted from the ground.
  - Using a predetermined test area, count the number of broken or damaged tiles to get a percentage of probable breakage.
  - The survey can be done with binoculars, and normally provides a reliable, rough estimate of how many tiles need to be replaced.
  - Deterioration in clay roofing tile is easy to spot.
  - It is almost invariably the result of water absorption in tiles that were not manufactured properly.
  - Most roofing tile failures can be traced to the use of inadequate raw materials or lack of proper time and care in the production process.
  - The inferior tiles that result have a tendency to absorb moisture.

- The moisture then expands and contracts in response to the extremes of the freeze and thaw cycle in the north and to heating by the sun in warmer climates.
- This process causes the tile body to flake and spall.
- The first sign of trouble is usually small chips of tile in gutters or around the foundation.
- Areas of discoloration, visible from a distance, might indicate that the internal body of the tile, which is lighter in color, is showing because the surface has chipped or flaked.
- The tile might appear fuzzy at the edges, and the shapes might be unclear.
- A closer inspection on the roof itself might reveal a crazed pattern of cracks in individual tile bodies.
- If such problems with spalling, cracking, etc., are widespread, the tile may in fact be deteriorated and in need of replacement.
- If there is any doubt, take representative samples and send them out to a reputable testing service.
- The tests look at the major factors in tile condition: pore structure, compressive strength, and water absorption rates.
- Good results on these tests indicate that the tile is a strong candidate for additional decades of useful life.
- When lifting and relaying tile roofs, weight can be a concern.
- If adequate scaffolding is available, or if the structure is deemed strong enough, the tiles can be lifted in sections and stacked near the work area.
- Common sense and experience are normally enough to let the roofer know whether this kind of loading might cause movement or possible collapse.
- When any doubt exists, consult with an engineer.
- If the roof or scaffolding is not strong enough, move the tile to the ground via a conveyor belt and develop a plan for restocking the roof with tile as needed.
- The actual removal of existing tile is straightforward:
  - The tile is simply lifted and rotated, which normally pries out the nail.
  - After the underlayment and deck are deemed satisfactory, the tiles can be relaid in the same manner as new tiles.

INSTALLING FIBER CEMENT SHAKES

A relative newcomer to concrete roofing is the fiber cement shake. This roofing product has the natural texture and tones of real cedar shakes with the added benefit of a Class A fire rating. Fiber cement shakes resist the damaging effects of sun, water, humidity, rot, fungus, and termites. They contain no asbestos, formaldehyde, or resins.

- For roof pitches 4/12 and greater, cement slates and shakes can be installed over spaced or solid sheathing (Figure 10.14).

- Install 18 inch wide, 30 pound underlayment, and the starter course at a 9 inch exposure.

- Then apply succeeding field courses at a typical 10 inch exposure.

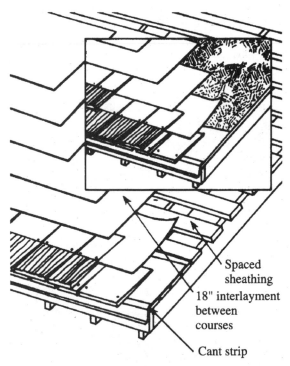

Spaced
sheathing
18" interlayment
between
courses

Cant strip

FIGURE 10.14 Installing flat cement tiles over spaced or solid sheathing on roofs with pitch of 4/12 and greater.

- Lap the underlayment over the ridge and hip to create a double layer.
- For a 1 inch stagger, use a 9 inch underlayment exposure.
- For roof pitches 3/12 and less than 4/12, lay shingles over solid sheathing.
- For 4/12 and greater pitches, install a 36 inch, 15 pound underlayment described previously.
- When roof pitches are under 3/12, install tile for appearance only over an approved sealed membrane, low slope roof system.
- In snow areas, an approved 36 inch snow and ice moisture barrier is recommended at the eave.

Fiber Cement Shinges

Use the following checklist at your jobsites. You can adapt the format to your company's protocol, then print it for the crew and send it with them.

- The treatment at ridge and hip in valleys and the flashing at chimneys, vents and so on, are handled in the same manner as that described for wooden shakes and slate in Chapters 8 and 9.
- Cement fiber shakes can be applied over an existing composition or wood shingled roof system, but the framing has to be sufficient and the existing system must smooth and uniform.
- Wind resistance performance applies only if fasteners penetrate the sheathing as specified.
- Structural evaluation and local building code approval is required.
- Longer fasteners are required when roofing over existing roof systems.

	INSTALLATION STEPS FOR FIBER CEMENT SHINGLES
	Install a 1/4 inch cant strip flush along the eave over the interlayment and beneath the starter course.
	Install the starter course face down with up to 1-1/2 inches overhang at eave and 1-1/2 to 2 inches at rake.
	Fasten the starter course with 1-3/4 inch corrosion resistant nails staples, or screws, located within 1 inch of the eave line.
	Interlap field shakes with interlayment to create an approximate 2 inch headlap and a typical 10.inch exposure to the weather.
	Install shakes with an approximate 1/2 inch keyway and 1-1/2 inch minimum sidelap. NOTE: Applications with more narrow keyways require additional material, which changes the appearance of the roof.
	Alternate the sequence of shake widths every third to fifth course to avoid a stair step pattern.
	Fasten the field shakes with either two 16 gauge 7/16 inch crown, 1-3/4 inch galvanized staples, or 13 gauge nails or screws. Locate fasteners approximately 10-1/2 inches above the butt, and 1 inch in from each side of the shake.
	NOTE: Do not walk on the smooth surface of installed field shakes.

FIGURE 10.15 Installing fiber cement shingles.

METAL ROOFS

The concept of using sheet metal as a roofing material dates back to the 12th century. Craftspeople in Germany, Scandinavia, and other countries connected small copper sheets for some of the first such applications. Lead was similarly utilized in 15th century Europe as artisans exploited its ductility and malleability.

Today's metal roofs are lightweight, yet extremely durable and cost effective over the long term. They have strong architectural appeal and can be adapted to suit a wide variety of historic and contemporary designs.

ADVANTAGES OF METAL ROOFS

- They install easily and quickly, yet are flexible enough to overcome unique or difficult situations.

- They are readily available throughout the United States in a broad range of colors and configurations.

Many technical improvements have been made in the basic, venerable metal roof system. Advances in coatings, treatments, and forming technologies, as well as major improvements in systems for seaming and anchoring, have combined with a renewed interest in historic architectural styles to make metal roofing systems a leading choice today.

The vast improvements in metal roofing, while offering many benefits, have not altered the need for great attention to detail during any installation.

TAKING ADVANTAGE OF METAL ROOFING

There are a number of reasons to use metal on commercial and residential buildings:

- Metal is a predictable and stable product.
- Unlike other single ply technologies like rubber, a metal panel is not affected by ultraviolet (UV) rays, which are one of the leading sources of roof degradation.
- Many standing-seam roof systems have earned a Class 90 wind-uplift rating (the highest in the industry) from Underwriters' Laboratories, Inc. (UL).
- Some systems also carry a Factory Mutual Class A fire rating as a result of their noncombustible surface.
- High ratings can help the building owner reduce insurance rates substantially.
- Metal roofing is extremely versatile.
- It is compatible with all types of building materials.
- Metal can be incorporated into all design concepts.
- Advances in metal coating and finishing technologies have significantly improved the life cycle of metal roofs.
- They have a long life span.
- Metal roofs are low maintenance.
- Metal roofing also is extremely puncture resistant.
- They are cost effective.
- Colors and finishes are available to suit any application, from earth tones to exotic tropical hues, from bare, unfinished surfaces to special coatings designed to resist corrosive atmospheres.
- These colors and finishes have warranties for as long as 20 years.
- The premium architectural finishes are highly resistant to fading and chalking.
- In addition to the 20 year warranties available on some panel finishes, the metal of the panel itself also carries warranties.

- Most manufacturers cover their panels against leaking due to penetration or rust for a period of 20 years.

- In the long run, for commercial, low slope construction, metal is the most economical material available.

- It is not the least expensive product initially, but when true life cycle cost comparisons are made, metal usually outperforms many other products that require periodic maintenance or replacement.

- On high pitch applications, architectural metal panels are favorably priced along with premium, fire retardant wooden shakes, clay tile, or slate.

- The use of metal roofing panels on a residence offers pride of ownership to the homeowner who wants to have something different from the neighbors.

- Metal panels are made to simulate clay tile, wooden shakes, and slate.

- Metal roofs are environment friendly, from installation through eventual tear off.

  - Many metal roofing products use recycled metal in their initial fabrication.

  - The building owner and contractor do not have to contend with external flames, hot kettles, or noxious fumes during installation.

  - After the roof has served its useful life, most of its metal is 100 percent recyclable, which eliminates expenses, liabilities, and concerns related to hazardous waste disposal.

METAL ROOFING

- Structural metal roofing:

  - Structural metal roofing has load carrying capabilities and does not require a substructure for support.

  - Structural metal roofing, depending on profile and material thickness, can be applied on very low slopes.

  - The industry minimum standard pitch for structural steel roofing systems is ¼ inch in 12 inches.

  - Other materials, such as aluminum, copper, and various alloys and plated composites, have structural limitations that need to be considered.

- Structural, standing-seam roofing is composed of interlocking panels that vary in width from 12 to 24 inches.
- The panels are designed to form a continuous waterproof membrane across the roof.
- The configuration of the panels is either a trapezoidal-shaped rib that is 2 to 3 inches high, or a square, vertical, narrow rib that is 2 to 3 inches high.
- Both of these basic panel configurations (Figure 11.1) rely on the shape of the panel to provide the structural capability to span structural members.
- The standard for structural panel free span is 5 feet.
- There are other factors to be considered, such as design load requirements and existing structural member locations, particularly in reroofing applications.
- These factors can alter the free-span capability, but 5 feet is the most common.
- Having a free-span capability means that no substructure is required to support the metal roofing panels.
- The structural, standing-seam panels provide support for construction personnel and maintenance traffic on the roof.
- In order for these structural, standing-seam panels to be waterproof, the sidelap joints and endlap joints are designed with continuous gasket seals.
- Most panel manufacturers install mastic in the ribs during the production process.
- The sidelaps are accomplished by mechanical seaming or snap-together seaming.
- These seaming methods compress the sealant into the sidelap joint and form a continuous seal along the rib.
- Endlaps are sealed with mastic or tape sealant placed between two layers of panels and then screwed together to form a compression seal.
- Structural, standing-seam panels are attached to the framing members with concealed fasteners and clips (Figure 11.2).
- The clips are designed to allow the panels to move back and forth in response to varying roof temperatures.
- Architectural metal roofing:
  - Architectural panels are nonstructural and must be supported by a substructure.

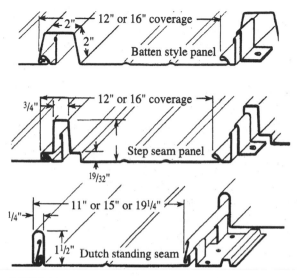

FIGURE 11.1 Basic metal panel configurations.

FIGURE 11.2 Typical concealed clip.

- In most cases, the substructural support is plywood decking.
- Metal decking or plywood treated with fire retardant is used when noncombustibility is required.
- Architectural panels, due to their design, are typically built to shed water, while structural standing seam panels are fabricated as true water barriers.

- The minimum roof slope for architectural panels is generally considered to be 3 inches in 12 inches.
- Roofing felts are usually installed between the panel and the substructure, to provide additional moisture protection.

Primary Categories of Architectural Metal Roofs

- Steep slope:
  - Also referred to as architectural metal, steep slope is used primarily for visual impact and typically requires a supporting deck with a minimum slope of 3/12.
  - The seams can vary, however, and they are usually hydrokinetic, or designed to shed water.
- Low slope:
  - Also referred to as structural, low-slope systems have a minimum slope of ¼ inch in 12 inches.
  - Typically, these systems can support their own weight without a deck.
  - The watertight, or hydrostatic, seams are designed to withstand water pressure, and the profiles and panels come in a wide variety of shapes and sizes.
  - In structural and architectural metal, thermal movement is accommodated in the field-seamed side joints and clip designs.
- Hybrid systems:
  - Hybrid systems are typically used in retrofit applications.
  - Hybrid systems combine both architectural and structural steel technologies to provide the building owner with long term roofing performance.
- A major advantage to reroofing with metal is the ability to create slope through the use of a sub-assembly system (Figure 11.3).
- When needed, a sloped sub-assembly made of light-steel structural elements can be attached to the existing roof surface to create a minimum ¼ to 12 pitch and ensure adequate drainage.
- The attic created by the subassembly affords a convenient area for adding insulation.

fastfacts

➤ *The crucial parts of any roof design are the details around projections and the perimeter of buildings.*

➤ *Metal details, regardless of the application or system used, can be extremely challenging.*

➤ *Aside from normal building movement, the joinery must be able to accommodate the expansion and contraction of the steel membrane.*

➤ *Simply put, metal roofing is less forgiving than most other single-ply, builtup (BUR), modified-bitumen (MBR), or asphalt-shingle technologies. When flashing a projection on a BUR, there can be three or four plies to provide a watertight seal.*

➤ *With metal, there is only one watertight seal.*

➤ *A roofer must be very precise, from the takeoff right through the final inspection.*

➤ *Because the roofer is dealing with large panels that cannot be easily patched or spliced to correct mistakes, errors can be expensive.*

FIGURE 11.3 Creating a slope/sub-assembly system roof frame.

SELECTING METAL ROOFING MATERIALS

Metal roofing materials include copper, aluminum, zinc, steel, and stainless steel. The following information describes some of the characteristics of metal roofs.

Copper

- Copper is expensive but remains the roof of choice on many important structures.
- It is simple to use.
- It offers natural beauty and extreme durability.
- The green patina on a copper roof is a weathering process that takes many years to complete.
- Chemical treatments can be used to accelerate the weathering process.
- The most common treatments use chloride-salt solutions or sulfate solutions that are sprayed, brushed, or sponged onto the roof.
- The success of the chemical treatment often depends on the weather at the time of application.
- One particular treatment, which uses an ammonium sulfate solution, requires a relative humidity of at least 80 percent while six to eight spray applications are made.
- A method that uses ammonium chloride requires dry weather for a 72 hour period while the treatment is on the roof.

fastfacts

➤ *The results obtained by any artificial weathering process vary greatly.*

➤ *Frequently, a convincing looking patina is achieved by artificial means, only to fade after a few years to one of the intermediate brown shades of the natural weathering process.*

➤ *Be very careful when choosing and applying patina accelerators on copper roof systems.*

Aluminum

- A mill finished aluminum surface rapidly develops a stable oxide coating that protects it from further corrosion.

- The oxide film is quite thin, on the order of a few millionths of an inch thick, and is clear and colorless.

- Because of this, the surface appearance does not change much with age.

- Bare aluminum surfaces are very good reflectors of radiant heat.

- A bright new aluminum surface reflects up to 98 percent of the radiant heat that strikes it.

- Even after weathering for years, the surface still reflects 85 to 95 percent of the radiant heat striking it.

- This means that a bare aluminum roof reflects most of the sun's radiant energy, thus holding down the temperature under the roof.

- Aluminizing is a process in which a layer of pure aluminum is deposited on steel.

- Aluminum coatings have good weathering characteristics and protect, by sacrificial corrosion, the base steel at breaks in the coating.

- Aluminum shingle systems come in several configurations, most of which have either 10 or 12 inch exposures.

- Aluminum shingles provide a realistic wood shake look because of their deep drawn graining and their formed-butt thicknesses, which vary by as much as $1\frac{1}{4}$ inches.

- Although the panel widths offered by different manufacturers range from 24 to 60 inches, most systems feature upturned and downturned top, bottom, and side flanges that permit the panels to be interlocked four ways.

Stainless Steel

- Stainless steel is an alloy that contains at least 11 percent chromium in combination with other alloying elements such as nickel, manganese, and molybdenum.

- The higher the chromium content in the alloy, the more resistant it is to corrosion.

- Some stainless steel alloys contain nearly 30 percent chromium.

- The type of stainless steel most commonly used for roofing products is Type 1818, which contains 18 percent chromium and 8 percent nickel as the main alloying elements.

EVALUATING METAL COATINGS

Galvanized, aluminum coated, and aluminum zinc coated sheet steels have contributed significantly to the growth of the standing-seam roof system. This has resulted in a number of quality products that perform essentially trouble free over the practical life of the roof.

- Standing seam metal roofing is made from cold-rolled, low carbon sheet steel.
- They are typically 0.019 to 0.024 inches thick.
- A hot dip metallic coating is added to provide the long term weather protection needed for atmospheric exposure.
- Zinc, aluminum, or a combination of both metals usually composes the metallic coating, which is metallurgically bonded to the steel substrate by passing the sheet steel through a molten bath.
- These coated sheet steels can be roll-formed easily into roofing panel profiles without the coating flaking, peeling, or powdering.
- Metallic coatings provide two types of corrosion protection, sacrificial and barrier.
- Sacrificial:
  - The coating oxidizes at a low rate to protect the base steel even at the uncoated, exposed edges.
- Barrier:
  - The coating acts as a wall between the base steel and the environment.
- Galvanized sheet steel has a zinc coating that utilizes both protective mechanisms.
- Sacrificial is the primary protection, and barrier is the secondary.
- The zinc coating is more reactive and sacrifices itself to prevent the steel substrate from corroding.
- Zinc coating weights for galvanized sheet are specified in American Society for Testing and Materials (ASTM) A525. Commercial roofing applications are typically G90, or a total of 0.90 ounce of zinc per square foot for both sides of the steel sheet.

- The coating measures slightly less than 0.001 inch thick per side and usually has a bold, spangled appearance.
- For painted applications, the spangle normally is subdued.
- Aluminum coated sheet steel relies almost solely on barrier protection.
- Sacrificial protection is provided only in the presence of chloride ions, such as in a marine environment.
- Aluminum coating weights are specified in ASTM A463.
- The minimum aluminum coating weight for roofing is 0.65 ounce per square foot total for both sides.
- The coating thickness is 0.0014 inch per side.
- The durability of aluminum coated sheets is less dependent on the severity of the environment than that of galvanized sheets.
- Aluminum coated sheets have a spangle-free or matte finish.
- Aluminum zinc coated steel is better known by its trade names:
  - Galvalume is approximately 55 percent aluminum and 45 percent zinc by weight.
  - Galfan is 95 percent zinc and 5 percent aluminum.
- Both materials provide sacrificial and barrier corrosion protection, although the higher the zinc content, the more sacrificial protection provided.
- As in the case of galvanized steel, the zinc-rich portion of the coating sacrifices itself to protect the steel substrate.
- The aluminum-rich part of the coating provides a long lasting barrier between the atmosphere and the base steel.
- Coating weights for Galvalume are specified in ASTM A792. Galfan, with coating weights similar to those for galvanized steel, is covered by ASTM A875.
- Galvalume and Galfan offer a corrosion resistant sheet that has the durability of aluminum coatings and the sacrificial protection of zinc coatings.
- The aluminum zinc metallic surface is less spangled than a galvanized sheet, but is still bright and reflective when unpainted.
- Galvalume, Galfan, and aluminum coated sheets can carry a limited 20 year warranty when used for roofing.

CHOOSING PAINTS AND LAMINATES

In these cases, organic coatings such as paints and laminates are applied over the metallic coating via coil coating, a process of adding an organic finish to a continuous strip of metal.

Paints

The most widely used paint systems for standing-seam roofs are polyesters, silicon polyesters, and fluorocarbons. Polyesters are organic polymers characterized by relatively hard and abrasion-resistant surfaces.

Silicon polyesters are organic polymers that offer exterior durability, principally, through chalk resistance and gloss retention. Fluorocarbons are vinyl polymers characterized by being both malleable and durable. They are also heat and chalk-resistant, and are good for color retention.

Laminates

Laminates are applied as a plastic film rather than a liquid, and are three times thicker than paints. They are adhesively bonded to the coil under pressure and heat during the coil-coating process. Plastic laminates prevent chalking, fading, peeling, and other forms of degradation. Acrylic and fluorocarbon laminates most commonly are used in roofing applications.

USING METAL WORKING TOOLS

Most roofing contractors have an array of more or less specialized metal working tools that range from hand-held shears, soldering irons, steel cutting saws, screw guns, and nibblers to sophisticated break presses, form flashings, seamers, and forming machines.

- Generally, an abrasive blade is not recommended for cutting panels.
- The heat it generates can burn away the galvanization and consequently leave the metal open to the development of rust.
- A nibbler is recommended for cutting panels.

fastfacts

➤ *Both laminate and paint finishes are nearly as protective as the metallic coatings beneath them.*

➤ *Both coatings enhance the appearance and increase corrosion resistance of the roof.*

➤ *The selection of metallic and organic coatings for application over steel substrates is best determined according to specific need.*

➤ *No single coating can be generally stated to be the best.*

➤ *While there are trends in the uses of these products, the smart choices in their use are governed primarily by the type of structure and environmental conditions.*

- Roll formers:
  - With a PRF machine, the roofing contractor is able to go directly to the jobsite to fabricate and custom fit the metal panels.
  - Once made, the panels are sent to the roof to be fitted next to the preceding form.
  - Some PRFs have optional panel profiles that can be changed out in a matter of a few hours.
  - PRF machines are expensive.
  - The metal roofing contractor must conduct a careful cost analysis between manufactured boxed panels and those made on PRF machines.
- Power seamers:
  - These small machines do the big job of forming the individual panels into a monolithic sheet.
  - The seam must be constructed appropriately to ensure a proper fit from panel to panel and to provide a watertight seal.
  - Power seamers are merely smaller versions of the larger PRF machines.
  - Powered by a heavy duty electric motor, power seamers do their job by locking onto the two standing seams of side-by-side metal panels.

- Using a series of driven rollers and idler rollers, the two metal panels are joined to form a lock.

- Power seamers can be fitted out to form several types of seams, single or double locks, tee locks, and batten seams.

- Some models are adaptable to more than one type of seam and can make a double lock in one pass.

- Depending on the gauge or type of material and the type of seam desired, power seamers do their job at a rate of 30 to 60 feet per minute.

- Once set, the power seamer works on its own, allowing the mechanic the luxury of attending to other tasks.

UNLOADING AND STORING METAL MATERIALS

To avoid damage, observe certain precautions when handling the roofing panels before installation.

- When unloading panels, take care not to dent or puncture them.

- If the material is unloaded with a crane, use a fabric sling and blocking to prevent crushing.

- Provide adequate support when bundles of long panels are offloaded from a truck.

- Support usually is required at the third points of a bundle.

fastfacts

➤ *The power seamer is a relatively simple device to operate, but it performs an important part of a standing-seam metal roofing job.*

➤ *In addition to greatly affecting the way the finished job looks, it ties the panels together and bonds them to the cleats that hold the roof to the deck.*

➤ *With the seams properly completed, the rest of the work has a tendency to take care of itself.*

- If the panels must be stored outdoors, place them off the ground on angled wooden blocking so that water can drain freely.
- Cover the stack of panels with a breathable tarp that keeps the rain off the panels and allows air to circulate, which lessens condensation buildup.
- White rust or staining tends to occur on tightly stacked panels between which moisture is trapped.
- Do not drag panels over one another or on the ground.
- Lift off and carry the bundle to avoid scratches that can cause the finish to fail prematurely.
- A common procedure during installation is to place bundles of panels at intervals on the roof purlins or deck.
- If the bundles are quite large, however, the roof structure can be overloaded in spots unless extra support is added.

FASTENING PANELS

- Metal roof systems use either exposed or concealed fasteners.
- Exposed fasteners are usually painted to match the color of the panels, while concealed fasteners are not.
- Most exposed fastener systems are quite similar in appearance and function, whereas the concealed fastener systems have a wide variety of panel profiles.

Exposed Fastener Systems

- The exposed fastener method is the easiest and probably least expensive way to secure a metal roof.
- The method might better be described as a lap-and-fasten system.
- The panels are lapped at the edges and a screw or nail is used to secure the joint.
- The fastener is driven through the high point of the lap with a washer to seal the hole.
- With exposed-fastener systems, a large number of holes must be created in the roof panels to fasten them to the purlins.
- Neoprene washers and other steps are taken to seal the fastener holes, but many potential leak sites remain.

- Temperature fluctuations also cause dimensional changes in the roof panels, making the panels slide relative to the purlins at points where the fasteners are not tight enough.

- This process can enlarge the fastener holes so much that the neoprene washers cannot provide a tight seal.

- Correct tightening of the fasteners is important to allow the neoprene washer to seal properly.

- An overly tight fastener squeezes the washer out of the joint or possibly tears the washer.

- In either case, a poor seal is created.

- A neoprene plug usually is installed ½ inch up the batten.

- An aluminum batten closure is slipped in and riveted to the side of the batten.

- Apply sealer to the inside of the batten before the closure is installed.

- If all fasteners are tightened so that the roof panels cannot slide relative to the purlins, then thermal expansion and contraction of the panels is absorbed by a process known as purlin roll.

- The purlins, typically Z-purlins, are of a light enough gauge to flex back and forth slightly as the roof panels expand and contract longitudinally.

- If purlins are inadvertently braced in a way that prevents them from rolling, repeated expansion and contraction of the panels can eventually snap off the fasteners or cause the fastener holes to leak.

- Different methods are used to keep water from creeping under the sidelap joint.

- Some manufacturers rely on factory or field applied sealant to create an impervious barrier.

- Others form an antisiphon or anticapillary groove in the lower panel of the lap.

- The groove creates a capillary break that prevents water from being drawn through the seam. In addition, the groove acts as a drain to carry away the water it has trapped.

Concealed-Fastener Systems

- As the name implies, the concealed-fastener system leaves no fasteners exposed to the elements on the longitudinal seams.

- Stamped metal clips are fastened to the substrate and hold the panels in place at the longitudinal seams.
- There are three types of clips:
  - Two-piece
  - Purlin-slip
  - Panel-slip
- There are two main types of concealed fastener systems:
  - Two-piece
  - Three-piece

Two-Piece Clips

- These clips are used with long, continuous panels that can undergo major thermal expansion and contraction.
- They consist of an upper part, which is clamped rigidly into the seam, and a lower part, which is fastened firmly to the substrate.
- The two pieces are connected by a slip joint that allows longitudinal movement of the panels. The ideal two-piece clip is self-centering.
- When it is installed, it is automatically set to allow equal amounts of movement in either direction.
- Problems can arise when long roof panels are installed in extremely hot or cold weather.
- Most of the thermal movement from that time on will be in one direction only.
- In these cases, the sliding portion of the clip may have to be installed off center to handle the expected thermal movement.

Purlin-Slip and Panel-Slip Clips

- These clips are used for short panel runs for which little thermal movement is expected.
- The two clips are basically the same.
- They differ in the way they are fastened.
- Purlin-slip clips allow for thermal movement in the loose connection between clip and substrate.
- Panel-slip clips are fastened tightly to the substrate, and thermal movement occurs when the panels slip relative to the clip.

- This slippage can cause some local damage to any sealant in the seam, but since the amount of thermal movement is small, the damage should be negligible.

- Regardless of which type of clip is used, there is some point along the slope at which the roof panels must be fixed to the substrate, or else the panels can walk their way down the slope with repeated thermal cycling.

- Possible fixing points are at the ridge, the eave, or some point near the middle of the slope.

- If the roof panels are fixed at the ridge or eave, flashing details are simplified at that point because there is no need to accommodate longitudinal movement of the roof panels.

- This means that the entire range of longitudinal thermal expansion and contraction must be dealt with at the opposite end, however, and movement can be considerable with very long slopes.

- A mid-slope fixing point can be desirable on very long slopes because it cuts in half the maximum amount of thermal expansion and contraction that must be accommodated.

- In this case, both the eave and the ridge flashing details must account for longitudinal movement of the roof panels, but again, the magnitude of the movement is less than would occur with the panels fixed at the ridge or eave alone.

Three-Piece Seams

- Components of a three-piece seam include the two adjacent panels and the seam cap.

- The three-piece seam has a variety of profiles, but all are merely variations of three basic approaches: raised, flush, and open seams.

- Three-piece seams have an advantage over two-piece seams when it is necessary to replace a damaged panel.

- A panel secured by a three-piece seam can be taken off by first removing the seam cap at both sides of the panel.

- The panel can then be lifted off and replaced with little or no damage to adjacent panels.

- The seam cap may not be reusable, depending on how it was originally applied, but a new seam cap is a minor expense.

- If a panel secured by a machine-seamed two-piece seam (Figure 11.4) needs to be replaced, there is a chance that one or both of the adjacent panels may be damaged in the removal process.

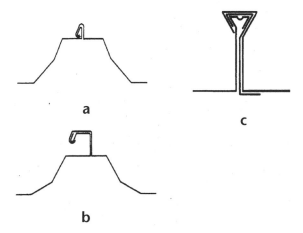

FIGURE 11.4 (a and b) Two-piece Type I raised-seam profiles. (c) Two-piece Type II.

FIGURE 11.5 Two-piece, Type I flush-seam profile.

- On the other hand, some of the snap together, two-piece seams appear to be very easy to separate, particularly the type shown in Figure 11.5.
- Some manufacturers provide a special tool for unlocking the seams on their panels.
- A disadvantage of the three-seam profile is that the seam cap is subject to damage or loss at the jobsite before it is applied to the seam.

APPLYING METAL ROOFING PANELS

The three stages of application are substrate preparation, panel membrane application, and terminations. Substrate preparation consists of erecting the purlin framework or solid deck and all other materials that go under the roof membrane. Panel membrane application involves laying and locking the roof panels. Terminations involve the installation of curbs, ridge caps, gutters, and any other finishing elements.

Preparing the Substrate

- Two different roofing situations can occur when preparing the substrate:
  - A new structure that requires a roof.
  - An existing structure that needs to be reroofed.
- The roof membrane that is to be recovered can be a BUR or metal type.
- Two different types of substrates on new structures:
  - Solid
  - Purlin
- The solid substrate can consist of wooden sheathing or metal decking.
- Rigid insulation can be placed on top of the sheathing or decking, or blanket insulation can be secured underneath.
- Certain types of roofing panels require the support of a solid substrate.
- A solid substrate provides structural bracing for the building and is occasionally used for that reason under roofing panels that do not actually require continuous support.
- The purlin substrate can be used only under structural-type panels that have spanning capabilities.
- The purlins, commonly light gauge C or Z bar steel members, are typically spaced 5 feet on center and span rafters.
- Blanket or rigid insulation can be laid over the purlins as desired.
- Consider the insulation's strength and support requirements when determining purlin spacing.
- A properly located vapor retardant is important in either type of substrate system to avoid problems with condensation in the insulation or on the underside of the roof panels.
- The most common reroofing job is to install a metal roof over a worn out BUR.

fastfacts

➤ *Old roof membrane can be left in place unless water has penetrated it and soaked the insulation below, in which case the roof membrane and insulation should be stripped away.*

➤ *Old insulation and roof membrane cannot be relied on to provide a firm, stable base for the new roof.*

➤ *In addition, any contained moisture could cause problems for the new roof system if the roof cavity is not vented properly.*

- Stripping off the old BUR adds considerably to the cost and labor of installing a new roof and negates one of the attractive features of a retrofit metal roof—the ability to install the metal roof on top of the old roof while causing minimal disturbance to inside activities.

- If the old roof membrane can be left in place, the loose ballast is usually removed to reduce the dead load.

- If the old roof is flat, a wooden or metal framework must be constructed to slope the new roof for drainage.

- Fasten the purlins to the framework and finish the roof in the normal way. If the old roof already has enough slope, it is not necessary to construct a framework for the purlins. They can be attached directly to the roof surface at the proper spacing. If a thick layer of insulation is used, the purlins might have to be raised up on blocking.

Adding the Panels

- Different roof panels form their sidelap seams using a variety of methods.

- The main difference in roof panel application methods is in how the panels are laid, from left to right, right to left, or across the slope of the roof.

- If the roof slope is short enough to be covered by a single panel length, the direction of application is arbitrary and can be left to the designer's and installer's judgment.

- If the roof slope is long enough to require two or more roof panels to be end lapped, the lay of the panel is usually dictated by the way in which the ends of the roof panels are prepared for the endlap.

- Most manufacturers prepare the panel ends that are to be lapped by either trimming away part of the seam on one panel or swaging or die setting the end of one or both panels.

- Factory preparation of the panels makes for a close fitting endlap and, consequently, a better seal.

- An added benefit is that field installers do not have to worry about trimming the panels.

- Once the direction of installation has been determined, the first roof panel or row of panels is laid at, or within, a few inches of the edge of the roof.

- The important step at this point is to be sure that the first panel or row of panels is laid exactly parallel to the roof.

- A slightly inaccurate alignment can be corrected as additional panels are laid, but the best method is to avoid misalignment in the first place.

- Some manufacturers offer a spacing gauge that can be used to make sure the roof panels are not inadvertently stretched or shortened in width as they are applied.

- Check panel alignment often as the installation progresses.

- It is not difficult to skew a relatively flexible metal panel that might be 50 or 60 feet long.

Sealing Terminations

Terminations are components such as ridge caps, curbs, flashings, gutters, and valleys that seal around openings or irregularities in the roof membrane and carry water to the building exterior.

Curbs And Pipe Flashing

- Terminations seal around penetrations through the roof panels.

- Curbs are used around larger, usually rectangular penetrations such as skylights, rooftop equipment, and ventilation ducts. Pipe flashings, as the name implies, are used to seal around pipe penetrations, such as a plumbing vent stack. Most pipes, ducts, and equipment items that penetrate the roof respond to thermal expansion and contraction. The curb or pipe flashing must allow for this relative movement while remaining weathertight.

- A skylight usually floats with the roof panels, so relative movement is not a problem with a skylight curb.

- An important feature for a wide curb is a diverter on the curb's upslope side.

- The diverter directs roof drainage around the curb rather than allowing it to build up against the curb.

- Center pipe and vent penetrations as closely as possible on the rib to allow maximum movement of the roof relative to the pipe.

- The rubber boot should provide a tight seal between the pipe and the sleeve.

- The location of a pipe penetration is not always the roofer's choice, so the pipe may not be centered in the sleeve.

- Many installers use a rubber-booted device to seal pipe penetrations.

- The rubber collar must be field cut to the proper diameter so that it seals snugly around the pipe.

- The hole in the panel must be large enough to allow the panel to move relative to the pipe.

- The pipe boot base should be 2 inches smaller than the distance between the standing seams.

- When the pipe penetration is at the ridge cap, the metal flashing is usually clamped and screwed to the ridge cap.

- The longitudinal thermal movement of the ridge cap is thus prevented at this point (Figure 11.6).

- The ridge cap should not buckle from thermal expansion and contraction as long as it is in sections no longer than 10 to 12 feet.

Flashing

- There are several ways to use a diverter with a curbed opening.

- Relative movement is possible between the curb and the roof, since the two are structurally independent of each other (Figure 11.7).

- The diverter directs water to either side of the curb.

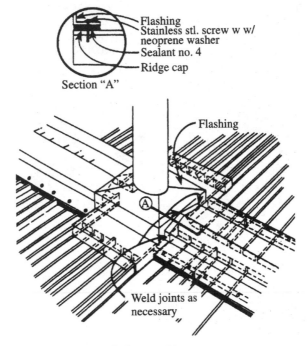

Flashing
Stainless stl. screw w w/
neoprene washer
Sealant no. 4
Ridge cap

Section "A"

Flashing

Ⓐ

Weld joints as
necessary "

FIGURE 11.6   Pipe flashing at ridge cap.

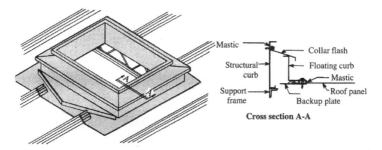

Mastic
Structural
curb
Support
frame

Collar flash
Floating curb
Mastic
Roof panel
Backup plate

Cross section A-A

FIGURE 11.7   Curbed opening with diverter that directs water to either side
of curb.

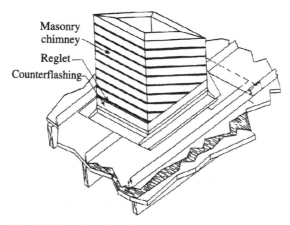

FIGURE 11.8 Use of reglets.

- The collar flashing does not overlap the floating curb by much in the vertical dimension.
- The installed height of the structural curb assembly is crucial to avoiding a gap where rain can enter.
- With masonry penetrations, such as the chimney in Figure 11.8, it would be wise to use reglets.
- A reglet is a slot cut in the chimney or concrete surface to hold the cap flashing.
- The reglet runs straight across the front and straight across the back, assuming there is no cricket.
- Use a diamond blade with a water spray attachment in a hand held circular saw, or a portable grinder with a masonry blade.
- A circular saw with a carbide masonry blade works, but not as fast as a grinder. In many cases, a cold chisel is really all that is needed.

Panel Endlaps

- These terminations are used when the roof slope is too long to be covered by one continuous length of roof panel.
- Some manufacturers specify that endlaps be staggered on adjacent rows of panels to minimize the number of extra plies of metal in the sidelap seam.

- In addition, one of the panels can be factory trimmed to reduce the extra plies. Other manufacturers specify that endlaps be aligned on adjacent rows of panels, rather than staggered. In this situation, one or more panel edges are usually trimmed and/or swaged at the factory to provide a smooth fit in the sidelap seam.

- On purlin-supported roofs, the endlap can occur either over a purlin, or where the lap is over the purlin to provide support.

- When an endlap occurs over a purlin, the fasteners are often driven into the purlin to draw the connection tight.

- Consequently, the roof panels cannot move to accommodate thermal expansion and contraction at that point.

- This condition must be kept in mind and the panels should be allowed to float at both the eave and the ridge to avoid buckling.

- The main concern when a lap joint occurs away from a purlin is whether the lap joint is strong enough to support occasional foot traffic.

Ridge Caps

- There are several different ridge-cap treatments possible.

- The typical ridge detail consists of pan closures, anchor strips, and a ridge cap.

- When designing the ridge, thermal movement must be taken into consideration.

- Depending on the length of the panels and where they are anchored, the style of the cap also comes into play.

- Simple ridge ventilation is acquired by using a screen with the step-up style ridge cap (Figure 11.9a).

- Never fasten the cap directly through the roof panel. Fasten the pan closure and the anchor strip to the batten (Figure 11.9b).

- The ridge cap then covers the anchor strip and is fastened to the strip between the bat tens.

- For thermal allowance, anchor the ridge cap on one side and allow the other side to float.

- For the float side, hook a return on the cap over a wider anchor strip. Both sides of the ridge cap should be the same size.

- When a neoprene strip is installed between the battens, the strips can be engaged in a U or J channel with cutouts over the batten.

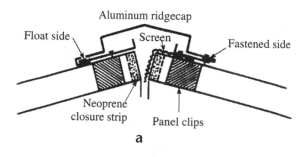

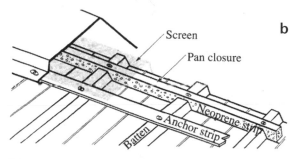

FIGURE 11.9 (a) Simple ridge ventilator system. (b) Pan closure and anchor strip.

- At this point, fasten them to the roof.
- Cover the installation with the ridge cap.
- For a shorter roof, a low ridge cap can be used (Figure 11.10).
- Attach the low ridge cap to the anchored end of the panel.
- The low ridge cap does not flex with the thermal movement.
- A float with a wider anchor strip can allow for thermal movement.
- Using wide low caps without step-ups is not recommended because they have a tendency to wave.
- Extend the roof panel all the way to the top of the roof when a low ridge cap is used.
- Short roof and hip caps can use the one-piece hip cap (Figure 11.11).
- Install the hip cap first, and then slip the panel into the channels.

- Install neoprene weather strips or similar closures to prevent water from blowing back under the joint.
- Pan closures at the ridge:
  - To turn up the pan on the ridge, slash the lock and batten and cut them 2 inches (Figure 11.12) to make the pan easier to bend.
  - When the pan is turned up the 2 inches, a tuck forms on the batten side, and the lock side is covered with the batten of the following panel.
  - Install a neoprene batten closure under the panel batten and an anchor strip across the length of the roof.
  - Install the ridge cap so that it covers the pan closure and finishes the roof ridge.

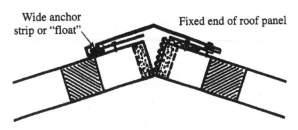

FIGURE 11.10 Low ridge cap.

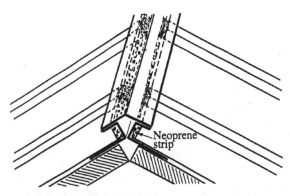

FIGURE 11.11 Roof and hip cap.

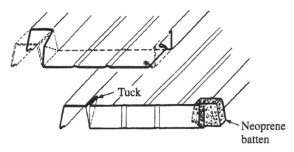

FIGURE 11.12 Pan closures at the ridge.

- Ridge cap overlap or splice joint:
  - A simple ridge cap can be overlapped.
  - The one-piece ridge cap or the step-up roof cap needs to be butted and spliced.
  - A splice joint can be installed under (Figure 11.13a) or over (Figure 11.13b) the roof cap.
  - Anchor the splice cover to one side of the joint only to allow for thermal movement.
  - Rake penetration between the batten and on the outside of the water barrier.
  - The cover can be smooth or an X can be field formed into the cover for a closer fit.
- Shop-formed ridge panel:
  - Roof panels are fixed at the ridge purlins, so panels must be allowed to float at the eaves.

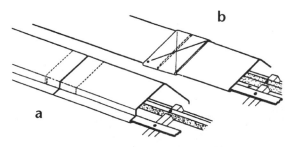

FIGURE 11.13 Ridge cap overlap or splice joint.

- The lap joints are not staggered on adjacent panels.
- Even though the edge of one panel is trimmed back, there is still one extra ply of metal in the seam, which can produce a poor seal.
- Field cut ridge panel:
  - These roof panels are fixed at the ridge purlins, so the panels must be allowed to float at the eaves.
  - The roof panel is trimmed where it meets the ridge section, so the seam cap does not have an extra ply of metal to enclose.
  - Lap joints are not staggered.
  - The vertical portion of the seam contains two extra plies of metal.
  - This design probably has no problems because of the seaming method.
  - It can be difficult, however, to seal the rib cut properly at the ridge.
- Field bent ridge panel—fix these roof panels at the ridge purlins.
  - Problems can occur if adjacent panels are allowed to move relative to one another near the ridge bend.
  - If the bend is created at the jobsite there also could be some difficulty creating the ridge bend at the proper location on each panel.
- Ridge cap with profile closure:
  - Allow the ridge cap to float longitudinally at the ridge.
  - Attach it in lengths of no more than 10 to 12 feet because of its longitudinal thermal expansion and contraction.
  - Provide extra support under the ridge cap to handle foot traffic and other heavy loads.
  - A profile closure is required to create the seal between the roof panels and the ridge cap.
  - The numerous through fasteners required to hold down the ridge cap create opportunities for leaks to form.

Hips

- Hip details are similar to ridge details.
- The panels leading to the hip are never cut to length.
- They are cut diagonally and, because of the variation of the angles, must be cut in the field.

- Allow enough length to cut the required angle.
- Take special care to match the ribs on both sides of the hip.
- When panels are field cut for rake edges, valleys, hips, etc., measure each panel before cutting to ensure the right fit and a straight alignment with the flashings.
- Due to various hip angles, the closures also are field cut and assembled.
- Install a U or J channel between the battens, with neoprene or another weather strip, and an anchor strip over the closures (Figure 11.14).
- Rivet the anchor strip to the batten.
- Hook the hip cap over one anchor strip and fasten it to the other anchor strip between battens.
- Install the hip cap before the ridge cap, since the thermal movement varies from one panel length to another.
- Give special attention to the design of the caps to allow for maximum movement.
- Anchor the panels leading to the hip cap at the eaves.

Valleys

- Valley flashings can be the most crucial flashings on a roof.
- Long panel runs emptying into a lengthy valley, especially on a low roof slope, require careful design and installation.

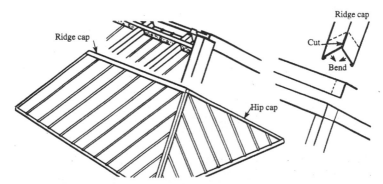

FIGURE 11.14 Closure of the pan on the hip.

- When designing roofs with valleys, consider using panels with vertical ribs.

- These panels are much easier to seal at a valley than trapezoidal-ribbed panels.

- Remember, in valley applications, the roof panel and closures are not of standard size because of the variation of the angle.

- The panels are cut diagonally and must be cut in the field to assure the right fit and angle.

- Use enough flashing to ensure the complete removal of all rain-water runoff, blowback, or overflow, as well as snow and ice.

- Figure 11.15 shows valley details with offset and integral cleats that eliminate exposed fasteners.

- A valley with integral cleats is one of the best designs because the valley trim is not punctured by fasteners, except at the top of each side.

- An inverted V in the middle of the valley flashing prevents cross flow and ensures flexing with thermal movement.

- This thermal movement varies from one end of the valley to the other.

- While the movement can be ignored on short roofs, on long lengths the movement needs to be considered when the cap or valley and flashing details are designed.

- To prevent relative movement between panels, give consideration to where the panels are anchored.

- Anchor valley panels at the ridge.

- Do not butt the roof panel against the V in the middle of the valley flashing.

- Leave an open space, or gully, to allow complete drainage of the panel.

- When using valley designs that require fasteners to puncture the valley trim, be sure that the fasteners are set in sealant.

- On long valleys, allow as much clearance as possible between the panel and the water diverter or the valley trim.

- This allows the valley to handle more water before it ponds onto the roof panels.

- Valleys that terminate before they reach the eave of a roof are especially difficult to install.

- The most common application of this detail is at a dormer at which one or more of the panel runs must have endlaps at the valley termination point.

- The lower panel laps under the valley trim, while the upper panel laps over the valley trim and is cut on the bevel to match the valley.

- Drive metal and neoprene plugs into the batten to prevent bugs or rodents from entering the panel.

- Because of the various panel angles, aluminum plugs must be field made and fitted.

- To hide the batten plug assembly, use the two-piece valley system and a cover plate.

- Install the cover plate over the V and reach across the gullies to the batten.

- Fasten the cover plate to the batten between the end of the panel and the neoprene plug.

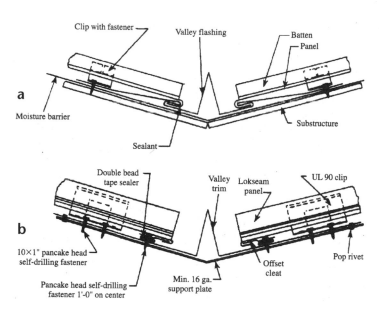

FIGURE 11.15 (a) Valley details with offset cleats. (b) Valley details with integrated cleats.

Rake Edge Or Endwall Flashing

- Use rake trim to seal the side of the roof to the building walls.
- On roofs designed to float or accommodate the thermal expansion or contraction of the roof panels, design the rake trim so that it floats.
- For roofs that do not need to float, use a fixed rake trim.
- Butt the panel batten up to the wall with the batten intact.
- Shape the flashing over the batten across the top of the wall and then over to the front and hook it into the fascia cleat.
- Cut the panel 1½ inches wider than the required width.
- Bend the 1½ inch piece upwards.
- Apply two bead strips of sealer and fasten an L channel between the sealer beads.
- Install the flashing over and onto the L channel.
- When a wall is much higher than the roof, use a separate flashing that is independent of the coping.
- Cut a reglet into the wall about 2 inches above and across the length of the roof.
- Apply sealer to the cut and slip the flashing into the reglet.
- Or, use a flashing with a 30 degree return.
- This acts as a caulking trough.
- Make sure that these types of applications are sealed thoroughly.
- Another way to cope with an unsightly wall is to install a metal panel soffit between the roof and the coping.

Parapet Flashing

- Generally, if the parapet wall is masonry, the parapet rake flashings are designed to float, even if the roof is fixed.
- This is due to the difference in rates of thermal movement between masonry and steel.
- Parapet high-eave flashings can be designed as fixed or floating.
- When dealing with masonry parapets, it is better to make a saw cut and insert a counterflashing than to use a surface mounted counterflashing.
- Surface mounted counterflashing can allow leaks due to cracks in the masonry parapet wall or a poor caulk job.

- In northern climates, snowdrifts can create additional waterproofing problems.
- Where the possibility of snowdrifts exists, carefully seal the parapet flash, rake, or high eave to the parapet wall.
- This is in addition to the shingle effect of the counterflashing overlapping the parapet flashing (Figure 11.16).

FINISHING CUSTOM DETAILS

- Several variables, such as the function and the appearance of the accessories, the length and anchoring of the panel, and environmental conditions, must be considered in the design and selection of these important customized components.
- Failure to assess these variables can hamper not only the effectiveness of the roof system, but also the aesthetic value of the building.
- To help find the optimum combination of aesthetics and function, this section illustrates several alternatives available to the building designer, contractor, or owner.
- It is recommended that all flashings, copings, etc., be installed as the roof installation progresses to avoid unnecessary traffic on the finished roof.

Fascia

- Different fascia designs require individual applications.
- Descriptions of the three most common types follow.
  - Simple fascia board: Use this type of installation only if water flow off the roof does not create any problems.
  - There should be adequate overhang and drainage to divert water away from the building.
  - Batten Strip: In this installation, cut the panel at the batten and lock and bend the pan in the field.
  - Cut and bend a separate batten strip to match the panel and install it over the batten by sealing and riveting it.
  - Kneecap and Batten Strip: A very striking design is made when the roof rib continues around the fascia and returns with the soffit to the building.

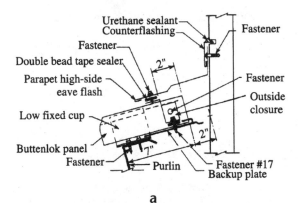

a

Urethane sealant
Counterflashing
Fastener
Fastener
Double bead tape sealer
Parapet high-side eave flash
Fastener
Low fixed cup
Outside closure
Buttenlok panel
Fastener
Purlin
Fastener #17
Backup plate
2"
2"
7"

Urethane sealant
Counterflashing
Parapet rake flash
Double bead tape sealer
Low-rake support
Parapet rake cleat
Parapet wall

b

FIGURE 11.16 (a) Floating high-eave parapet. (b) Fixed-eave parapet.

- Cut and bend the panel at two places. Use a one-piece batten strip to cover both knees.
- The top pieces always overlap the lower piece to prevent water intrusion.

Wall and Peak Cap

- This cap is similar to the ridge cap, but converts to a fascia board on the front.
- If the roof panel is anchored on the top, thermal movement is negligible.
- When the panel is anchored on the cave, thermal movement needs to be considered when the cap detail is designed.
- Install the roof panel with a closure and anchor strip prior to installing the cap.
- Slip the wall panels into the channel of the fascia trim.
- For shorter roof lengths or a shed roof, use a one-piece installation.
- Cut at the batten and lock.
- Bend at the pan.
- Install a kneecap cover over the batten.

Eave Fascia With Trim Strip

- Batten closures are often used along with neoprene plugs to close the roof section.
- Install the flashing first, and then proceed with the roof panel and vertical panel to align the ribs.
- Use bottom nose trim and soffit panels (Figure 11.17).
- Install the soffit panels onto the trusses, which are made individually and connected with the installation rail, or hat section.
- This creates a horizontal soffit.
- Use a conversion trim on this application.
- The variation in the angle of the roof makes the conversion trim a custom-formed trim piece.
- Install neoprene closures in the channels to prevent blowback.
- Be sure to line up the ribs of the wall panel with the ribs of the roof panel.

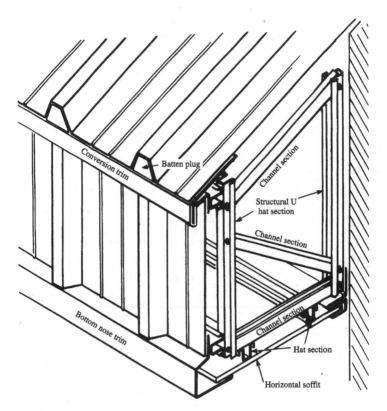

FIGURE 11.17 Eave and fascia detail with trim stop.

Diagonal Edges

- Due to variations in roof angles, the panels must be field measured and field cut.
- Make sure enough length is available to cut the required angle.
- Verify the length and angle of each panel for proper fit and alignment with the roof and flashing or trim.
- Install wide flashing under the roof panel and extend it to form a drip edge or fascia board.
- Install neoprene plugs in the batten and sealer used under the pan.
- Because of the angle of the batten, the aluminum closures need to be field formed to assure an accurate fit.

- The trim is installed over the batten and extended past the drip edge to allow for complete drainage off the pan fastened onto the battens.

- In both applications, water is allowed to fall freely from that part alignment with the roof and flashing or trim.

- Install wide flashing under the roof panel and extend it to form a drip edge or fascia board.

- Install neoprene plugs in the batten and sealer used under the pan.

- Because of the angle of the batten, the aluminum closures need to be field formed to assure an accurate fit.

- The trim is installed over the batten and extended past the drip edge to allow for complete drainage off the pan fastened onto the battens.

- In both applications, water is allowed to fall freely from that part of the roof.

- If draining to grade causes a problem, consider installing a gutter.

Building Expansion

- To allow for building expansion, the roof needs an expansion cap or crown.

- Fasten a simple crown to both sides of the expansion joint.

- Fasten the cap to one side and allow the other side to float.

- Work the panels outward from the expansion joint.

- Install a starter strip on both sides of the expansion joint.

- Allow enough space between the starter strips to hook the panels into place.

- An alternative method is to install a flat, or float-anchor, strip on one of the battens and hook the expansion cap into the float.

- Then fasten it to the other batten.

- Or, fasten the crown to both battens so that the V in the crown flexes with the thermal movement.

- Stop the roof panels at the expansion joint.

- If the last panel extends past the expansion joint, cut the panel 2 inches longer than the available space.

- Bend those 2 inches upward for extra strength.

- If the panel fits flush to the expansion joint, do not cut or bend it. Install a starter strip and continue with the next panel.

- For a float anchor, apply a Z-bar onto the cut panel and install a float-anchor strip to the batten of the next panel.

- Hook the expansion cap into the float and fasten it to the Z-bar.

- With the crown application, use an L channel on the cut panel and fasten the crown to the side of the batten and the L channel.

Gutters

- With metal roofing, it is possible to create two functional gutter systems that are architecturally aesthetic and not easily attainable with other roofing materials.

- Hidden gutters: This gutter system becomes an integral part of the roof. For aesthetic reasons, the gutter can be built into a facade. The downspout or leader can be hidden in a column or the wall.

- Recessed gutters: The recessed gutter system allows the gutter to be installed into the roof, either in the overhang area or within the wall.

- This type of installation has to be considered from the start and the structural steel needs to be designed accordingly.

- Often, a single ply roof membrane is used as the gutter liner.

REPAIRING AND CARING FOR METAL ROOFS

- A metal roof needs occasional care and maintenance to ensure that it lasts as long as possible.

- Most manufacturers recommend that the roof be washed annually with a strong stream of clear water to remove atmospheric dirt that can make the finish appear dull.

- This is a solid ancillary business.

- The roofer is used to heights, the equipment is relatively inexpensive, and the in-and-out nature of the work is excellent.

- Barring unforeseen damage, a yearly washing is all the maintenance the metal roof requires for many years.

- Occasionally, a roof is damaged and one or more panels need to be replaced.

- This procedure requires the opening of two or more sidelap seams so that the damaged panel(s) can be removed and replaced. In this situation, panels secured by three-piece seams are advantageous because adjacent, undamaged panels are disturbed minimally, if at all.

- Cut the seam cap for the three-piece seam along its entire length with an air chisel or other equipment and then lift out the damaged panel and lay a new panel in its place.
- Apply a new seam cap to complete the procedure.
- This can be a part of the cleaning service.
- Panels secured by two-piece seams, particularly those that have been machine seamed on the roof, usually are more difficult to remove without disturbing adjacent undamaged panels.
- Some manufacturers supply special tools that can be used to pry open the seams.
- Once the damaged panel is removed, lay the new panel in place and reseal the seams.

Recoating Options

- Recoating a corroded or badly weathered metal roof can be an economical alternative to total replacement.
- In general, the recoating procedure requires that any peeling paint and loose, heavy corrosion be removed by either wire brushing or sandblasting.
- The entire surface of the roof should then be cleaned thoroughly.
- Use a cleaning solution appropriate for the original finish, and allow the roof to dry completely.
- Always spot prime bare metal and tight corrosion.
- Some manufacturers require that the rest of the roof be primed as well.
- Then apply the finish coat at the manufacturer's recommended coverage rate.
- Although coatings such as urethanes and fluorocarbon polymers often are used in recoating applications, it should be pointed out that these formulations are not the standard factory applied coatings.
- Coatings that are intended to be factory applied normally require oven baking to cure properly.
- These coatings would fail if they were applied in the field.

- Field applied coatings are specially formulated for air curing.

- Manufacturers of these urethanes and fluorocarbons claim that life spans of the field applied coatings approach those of the factory applied, oven baked coatings.

- As metal roofs see more common use, the need to reroof structures with worn out metal roofs becomes more common.

- Often, the old metal roof becomes the substrate for the new metal roof.

- This situation raises questions about the compatibility of materials used on the old and new roof systems and the old roof's structural integrity.

- As explained in Chapter 7, galvanic corrosion occurs when two dissimilar metals come in contact with each other in the presence of an electrolyte.

- Moisture that might condense between the old and new roofs could act as the electrolyte and, if the metals are not compatible, cause hidden galvanic corrosion of the support clips or old roof.

- Moreover, a badly corroded or deteriorated metal roof might need to be removed entirely because it cannot provide a structurally sound surface for the new roof.

- If the old roof covering is left in place, check to see how it is attached to the building.

- The new roof cannot be expected to achieve a high wind uplift rating if the original roof is not able to achieve the same rating.

- If necessary, perform tests on the old roof to determine the pull-out strength of its fasteners.

Minimizing Oil Canning

- Oil canning is defined as a perceived waviness in the flat roof areas.

- It is an aesthetic problem that must be addressed if it is so noticeable that it distracts from the appearance of the building.

- Normally, structural integrity is not affected; but structural must be reviewed if the distortion is of an extreme nature.

- Since many uncontrollable factors are involved, no manufacturer can realistically assure the total elimination of oil canning.

- With careful attention to the production and selection of material, to the panel design, and to the installation practice, oil canning can be effectively minimized.

- Take the following precautions to control oil canning:
- Coil:
  - Tension leveling, a process whereby the metal is stretched in coil form, provides a flatter surface that is less prone to oil canning.
  - Also, the heavier the metal gauge, the less likely the product is to wave.
  - The possibility of oil canning can be reduced by ordering tension leveled material and specifying a minimal amount of allowable camber in the coil.
- Design:
  - The addition of stiffening beads breaks up the flat surface and makes oil canning less apparent.
  - Embossing also helps hide surface waviness in the metal.
  - The selection of low gloss coatings and light colors also tends to minimize the visual effect of oil canning.
- Installation:
  - More stringent specifications regarding the alignment of the supporting structure focus attention on this critical aspect.
  - Instructions regarding proper handling, spacing, and fastening should be a part of the manufacturer's delivery packet.

PLASTIC PANELS

- While not metal, this newest roofing material is installed in basically the same way as metal panels, and therefore the method of installation is included in this chapter.
- The thermoplastic roof panel is available in wooden shake and slate facsimiles, or in tile configurations.
- These panels have a better flame retardant value than the wooden shake.
- Plastic panels do not rot, curl, grow algae or moss, break, crack or split, or absorb moisture.
- Systems made from thermoplastic resin are not affected by salt spray or by acidic environments.
- They are virtually indestructible, since they are rust and termite-proof.

- The tile and slate configurations do not require structural reinforcement.
- Temperature fluctuations do not affect the performance of the panel.
- Its strength, even in sub-zero temperatures, defies hailstorms.
- Workers can walk on the roof without fear of denting, cracking, or breaking it.
- The lightweight of the roof panel makes it an ideal reroof product.
- Removal of the existing roof material is not required in many cases.
- This constitutes a tremendous savings of labor and disposal costs.
- Where there is an uneven roof in place, furring out to create an even surface might be required.
- Installing the panel over concrete or clay tiles is not recommended.
- The panels must be installed on, and the fasteners should penetrate into, a solid base for good anchorage.
- The panels are hooked to each other with overlapping top and sides, and should be installed in a staggered fashion for a more natural look.

Working with the Panels

- Recommended fasteners are self tapping screws, wood screws, or nails, depending on the application and the substructure.
- Take care to place the fasteners in the nailing slots.
- When a nail slot is not available, predrill a hole.
- The roof panel, which is molded of resin, can be cut with any circular or hand saw.
- Use a fine toothed blade with 10 to 18 teeth per inch.
- The wooden shake configuration generally has three courses of shake design per panel, is approximately 8 inches high per module, and has varied shake width.
- The tile configuration usually has two courses and the scallops that represent the Spanish tile have an 8 inch wide repeat.
- The country slate configuration generally has four courses and each slate is 6 inches high and approximately 8 inches wide.
- Side overlaps are staggered to achieve a natural look.

Typical Installation Guidelines

• Thermoplastic roof panels must be installed from left to right.

• The minimum recommended pitch is 3/12, and fasteners should be of sufficient length and pullout strength to meet local codes.

• At least three or four fasteners per panel are recommended.

• The thermoplastic roofing system is designed to be installed over solid sheathing.

• The roofing can be installed over low profile roofs, like asphalt.

• The recommended underlayment is equal to or better than No. 30 felt.

• Do not allow water to drain under the roof panels.

Panel Installation

• To ensure a level installation, snap a chalkline from the eaves of the roof.

• Install a ¼ inch J channel along the chalkline and fasten every 10 inches.

• Slide panel tabs into the J channel.

• Make sure the tabs are engaged into the J channel.

• Fasten the panel through the nailing slots.

• Do not overdrive the fasteners.

• Use fasteners that are long enough to penetrate the structural members or solid underlayment.

• Extend the panel ½ inch beyond the edge of the roof at the eaves.

• Cut the steps from the panel to assure a straight line up the rake edge.

• The panels can be cut with a fine-toothed saw blade.

• Install panel seams in a staggered fashion.

• What is cut from the end of the course can be used to start the next course.

• This minimizes waste at the beginning and end of each row.

• Fasten the panel at the end.

• Cover the end with the fastener with the rake edge.

• Install the rake edge as work progresses up the roof.

• The rake edge can be either metal or a specially designed rake edge.

Hips and Ridge

- Use a hip or ridge cap to cover the edge.
- Install the panels to the roof edge and use a metal strip to cover the cap.
- Or use a specially designed hip or ridge cap.

Sidewall Flashing

- Use metal for the sidewall flashing.
- Make sure there is positive drainage away from the wall and down the roof over the panel.

Valley Detail

- Use a metal valley flashing with generous side flanges.
- Form an inverted V in the center of the valley sheet.
- Attach the panels to the inverted V.
- When installing dormer or mid-roof valleys, make sure that there is a positive drainage of water onto the lower roof section.
- Overlap the lower panel with the valley sheet, and cut and install the panels to fit the angle.

Roof Penetrations

- A standard vent-pipe boot is available for most penetrations.
- Follow the manufacturer's installation instructions.
- Other roof penetrations might need to be flashed out according to field requirements.
- Use one-ply roof material or other weatherstripping to seal around penetrations.
- For a chimney, use a camelback, cricket-and-step, and/or reglet flashing, depending on its placement on the roof.
- All flashings must have positive drainage over the roof panels.

Apron Flashing

- When the panel fits to the wall, install the apron flashing with a reglet or fit it under the wall cover or windowsill.

- If the panel is too long, cut the extra material, install the panel, and cover it with the apron flashing.

Installing Plastic on New Roofs

- For new roofs, plywood is the recommended underlayment.
- Apply No. 30 felt paper or another specified vapor barrier.
- Install the panels as recommended and previously described.

Reroofing with Plastic

- Solve any existing problems with the old roof before installing the new roof.
- Pretreat known leaks and replace decayed underlayment or structural members.
- If the roof is in sound condition and relatively even, install the panel directly over the old roof.
- The thermoplastic panel is less than 70 pounds per 100 square feet.
- Make sure the fasteners penetrate the solid underlayment.

Cleaning Plastic Roofs

- Do not clean plastic panels with solvents such as alcohol, ketone, or aromatics.
- These solvents soften the finish.
- If necessary, wash with a mild detergent and water.
- Rinse with clear water.
- When contracting for any roof work that is new to your company, make certain that all design elements and manufacturer's instructions have been reviewed with care.
- Talk with the manufacturer about nearby installations and all tricks of the trade.

12

REROOFING

In contrast to new construction bids, in the reroofing market the lowest price is not always the key to success. Quality, service, and long term performance are the factors that attract reroofing business. Many building owners rely on referrals when searching for a roofing contractor. As a result, it is essential for the roofer to establish strong customer relations in his or her locale. Two imperatives for profitable reroofing are that you develop an excellent team for the work, and that you are consistent in asking existing clients to spread the word.

- Reroofing is always needed because old roofs are always wearing out and they need to be replaced.

- This is an excellent business medium for developing an ongoing cash flow.

- It is important to remember that not all roofing systems are well suited to reroofing projects.

- Simply because a contractor has experience with a few systems does not mean that those products can be installed on every job.

- When the smart roofer runs into invitations to bid on work that is not suited to the company, he or she is ready to walk away while guiding the prospective client to a steadfast contractor in the area who will do a good job.

- The proactive benefits in this decision are that he or she knows which jobs make the company money and does not take on work that can cause negative cash flow.
- Walking away from work is very important.
- The prospect will notice this integrity, and you can ask him or her for the names of friends who have roofs on which you know you will turn profits.
- You can also work out a marketing trade with roofing contractors in your vicinity who install roof systems that you do not want to incorporate into your business.

CHECKING THE OLD WORK

The principles for reroofing are essentially no different from those for new construction. However, there are additional factors that must be taken into consideration, such as what caused the previous roof to deteriorate and the extent of any damage that resulted from that deterioration.

- If either the insulation or the membrane is wet, they must be removed.
- Perform a thorough inspection of the roof to determine the requirements for rebuilding the structure.
- Do not automatically assume that a reroofing job requires a complete tear-off.
- Just because the shingles are curled at the edges, or because most of the granules are gone and the shingles are brittle, does not mean that the roof has failed. Be sure to discuss any weaknesses of the roof with the owner.
- Explain what life the system has left and sign a maintenance contract; this will probably help secure the reroofing for your firm.
- Check the sheathing and insulation from underneath the roof for signs of moisture penetration.
- Inspect the roof to find out how many roofs are currently on the structure.
- If the roof has failed and only one or two roofs are currently in place, a new roof might be placed directly over the existing roof.
- This assessment depends on the structure and the type of existing roofing material.

- Take into account the function of the building and the building's need for roof insulation and ventilation.

- Incorporate these findings in your reroofing bid when appropriate.

- They are essential components of a good roofing system, and they contribute to the bottom line.

- Evaluate the need to create a sloped roof if none exists.

- Check all masonry, stucco, copings, etc., for needed repairs that might be pertinent to the total waterproofing of the building.

- When a tear-off is required, address the job as professionally as possible.

- Take all necessary precautions to ensure that no collateral damage is done to the property through the careless removal of the current roofing system.

- Plan the most unobtrusive route for getting heavy equipment onto and off the property and into an effective working position.

- Be aware of the occupants of the building and their needs, as well as the roofer's requirements, including access to utilities and debris removal.

- Make certain that you carry the policies of the preroofing conference into all parts of your business.

- The reroofing meeting may not be as long or as intense as the meetings for new work. Many of the design elements will be dictated by the existing roof and structure. The meeting should cover all questions and details and is an excellent opportunity to sell extras. Figure 12.1 illustrates factors to consider for reroofing.

ROOF DRAIN SYSTEMS

More and more roofing contractors are installing roof drain systems. Drain systems are the single most vulnerable area for leakage on the roof, and while the contractor might have had nothing to do with their installation, they are high on his or her callback list.

- The architect or specifier usually designs the drainage system by taking into account such factors as the area to be drained (the size of gutters, downspouts, and outlets), the slope of the roof, the type of building, and the appearance.

- The design capacity for a roof system depends on the quantity of water that needs to be handled.

✓	Item
	Which roofing products are best suited to individual reroofing jobs?
	Which products offer the best profit margins?
	The customer's interests and desires concerning the property.
	The type and location of the property.
	The surrounding environment.
	How the customer can contribute to your marketing needs.
	Sell extras: gutters, ground drains, etc.
	Sell an extended warranty for mutual protection.
	Sell a maintenance plan to the client.

FIGURE 12.1 Factors to consider when reroofing.

- The quantity of water, in turn, depends on the roof area, slope, and rainfall intensity.
- When evaluating the roof area, it must be remembered that rain does not necessarily fall vertically and that maximum conditions exist only when rain falls perpendicular to a surface.
- Since the roof area increases as the pitch increases, it is not advisable to use the plan area of a pitched roof to calculate drainage.
- Though it is imperative that drains be designed for a capacity that will fully handle the "100-year rain," it is important to keep calculations accurate.
- Experience has taught us that using the true area of a pitched roof often leads to oversized gutters, downspouts, and drains.
- To determine the design area for a pitched roof, consult Table 12.1.
- When sizing gutters, the following factors must be considered: spacing and size of the outlet openings, roof slope, gutter style, and expansion-joint location.
- Since no gutters are effective for their full depth and width, the gutter must be designed so that water from a steep roof does not, by its own velocity, spill over the front and rear edges.

TABLE 12.1 Design of Gutter Systems

Areas for pitched roofs	
Pitch	**Factor***
Level to 3 in./ft	1.00
4 to 5 in./ft	1.05
6 to 8 in./ft	1.10
9 to 11 in./ft	1.20
12 in./ft	1.30

*To determine the design area, multiply the plan area by this factor.

- This can result in numerous types of damage to the building: scouring of siding, rot at rafters, and water intrusion below the gutters.

- The location of gutter expansion joints is important because water cannot flow past an expansion joint.

- If it backs up at joints and overflows, the same damages are apt to occur as with overflow caused by improper size.

Types and Sizes Of Gutters

- The size of half-round gutters is directly related to the downspout size.

- If the downspout spacing is 20 feet or less, the gutter size can be the same as the downspout size, but not less than 4 inches.

- If the downspout spacing is between 20 and 50 feet, add 1 inch to the downspout size for the gutter.

- Although draining more than 50 feet of gutter into one downspout is not usually recommended, whenever it is necessary, the gutter should be 2 inches larger than the downspout.

- The required sizes of other leaders can be determined by finding the semicircular or rectangular area that most closely fits the irregular cross section.

- The depth of a gutter should not be less than half or more than three-fourths of its width.

- Gutter sizes are usually expressed in width only.

- Half round gutters are economical and highly efficient.
- They are commonly used at eaves and troughs, and less often as built-in gutters.
- Other common types of gutters:
  - Pole gutters (Figure 12.2).
  - Molded gutters (Figure 12.3).
  - Builtin gutters (Figure 12.4).
  - Flat-roof drains (Figure 12.5).

Sizing Outlets

- Outlets or leaders (Figure 12.6) should be elliptical or rectangular in plan, with the longer dimension in the direction of gutter flow.
- The longer dimension should be the same as the gutter width and the shorter dimension about two-thirds of the gutter width (Table 12.2).
- Modern, factory-made gutters are wooden, metal, or plastic.
- At one time, wooden gutters were the favorite for residential drainage systems, but their popularity has decreased rapidly in the last few years. This is because of the weight, cumbersome nature, and cost of these systems, as well as the fact that wooden gutters require continual maintenance and applications of wood preservative to prevent rot.
- As a result, metal and plastic gutters are most frequently used for residential drainage replacement jobs.

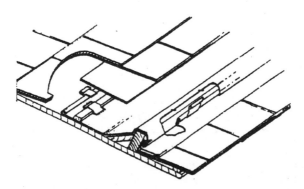

FIGURE 12.2 Pole gutter.

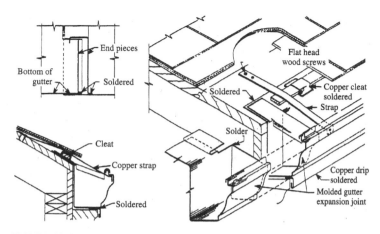

FIGURE 12.3 Molded gutter.

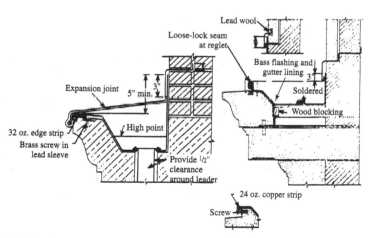

FIGURE 12.4 Built-in gutter.

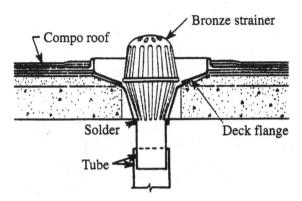

FIGURE 12.5 Typical flat roof drain.

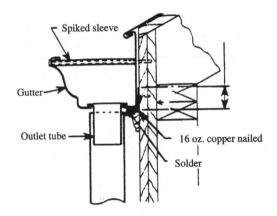

FIGURE 12.6 Outlet of a leader.

Metal Gutters

There are four types of metal gutters: galvanized steel, enameled galvanized, aluminum, and enameled aluminum. Figure 12.7 shows the various shapes of metal gutters.

- The least expensive metal gutters are typically plain, galvanized steel.
- They must be primed with a rust inhibitor and then painted.

TABLE 12.2 Dimensions of Standard Leaders

Type	Area (sq. in.)	Leader sizes (in.)
Plain round	7.07	3
	12.57	4
	19.63	5
	28.27	6
Corrugated round	5.94	3
	11.04	4
	17.72	5
	25.97	6
Polygon octagonal	6.36	3
	11.30	4
	17.65	5
	25.40	6
Square corrugated	3.80	$1^3/4 \times 2^1/4$ (2)
	7.73	$2^3/8 \times 3^1/4$ (3)
	11.70	$2^3/4 \times 4^1/4$ (4)
	18.75	$3^3/4 \times 5$ (5)

- Care must be taken to treat all exposed areas in order to prevent corrosion.

- Aluminum gutters do not rust when left unfinished, since they have their own natural protective coating.

- But certain chemical reactions do cause the deterioration of aluminum, and the installer should clarify all areas of vulnerability with the manufacturer and explain these thoroughly to the owner. This may be an excellent opportunity to sell a maintenance contract.

- Baked-on enamel finishes on products permit color matching without the hassles of painting.

- Keep in mind, however, that steel gutters rust if their enameled surface is scratched and the metal exposed.

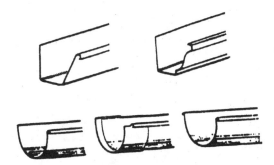

FIGURE 12.7 Various metal gutter shapes.

- Also, remember never to mix aluminum and steel gutters, since contact between the two metals starts electrolytic action and corrosion.
- Plastic gutters do not corrode, and the colors of the new products tend to be long lasting.
- The major problem with plastic gutters is their high coefficient of thermal expansion, which means that if they are not properly installed, they can buckle in hot weather.

Materials Take-offs

- To determine the amount of material needed, measure the length of the gutters around the house.
- Most gutters are available in 10 foot lengths.
- Figure how many multiples of 10 feet are required.
- Some manufacturers produce longer lengths of gutters, but these are more difficult to handle.
- Once you figure out the lengths of gutters and downspouts needed, plus the various accessories such as elbows, corners, and hangers, order the parts necessary for the installation.

Installing Metal Gutters

- Metal gutter is often installed with long gutter spikes that pass through the sides of the top of the gutter trough and into the fascia board (see Figure 12.8).

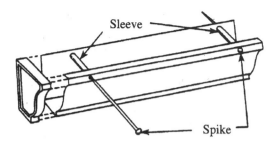

FIGURE 12.8 Installing metal gutters.

- A metal tube or sleeve, called a ferrule, mounts around the spike and maintains the spacing of the gutter sides.
- Locate these spikes 24 to 30 inches apart, depending on snow conditions.
- Two other hanging methods can be used for metal gutters:
  - Sickle-shaped hangers can be fastened to the fascia boards with spacing at approximately 30 inches.
  - The gutters are laid on top of the hanger (Figure 12.9).
  - This method eliminates the need to drill holes in the gutter and thus is easier than spiking in place.
  - The sickle shaped hangers are more expensive than spikes.
  - Strap hangers have flanges that are nailed in place under the roofing material (Figure 12.10).
  - This type of hanger is best used on new work, since there is always a chance of damaging the existing roofing when it is pried up to install the hanger.
- Elbows usually connect outlet tubes and collector spouts.
- Squeeze the end of the downspout and insert it into the large end of the elbow.
- Downspouts are fastened to the wall with downspout straps.
- Be sure the straps are long enough to anchor the downspout securely to the wall.
- If the gap between the wall and the downspout is too wide, shim the downspout strap 1 inch away from the wall with a product that is weather resistant or waterproof.

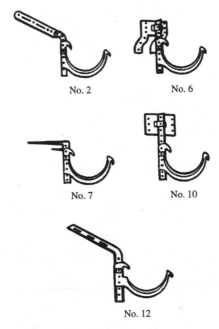

No. 2   No. 6

No. 7   No. 10

No. 12

FIGURE 12.9 Sickle shaped hangers.

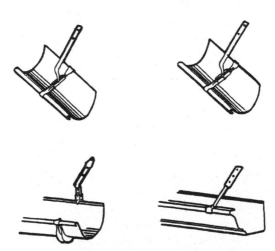

FIGURE 12.10 Strap hangers.

fastfacts

➤ *Draining water away from foundations is very important. Foundation problems are common triggers of very complex and expensive litigation.*

➤ *There are several ways to achieve this, but the most popular is to use a concrete splash block directly under the downspout, which runs the water toward the driveway or a similar draining surface*

➤ *Another common method is to connect the downspout directly to an underground line that leads to the storm sewer system.*

- Use two straps on each 10 foot length of downspout and three straps on two joined 10 foot sections.
- To enable downspouts to perform their maximum service, install them so that they carry water as far away from the building's foundation line as possible.

INSTALLING PLASTIC GUTTERS

- To install the gutter hangers, line them up with the chalkline and nail them at 18 inch centers.
- Be sure the section of each hanger that comes in contact with the fascia is perpendicular to the chalkline.
- Use only noncorrosive screws or nails.
- Locate hangers so that when the gutter is snapped in place, it catches the rain as it runs off the roof.
- For installations where there is no fascia board, use special gutter hanger rods.
- Nail these rods to the roof edge and bend them to the desired angle.
- Use collector boxes or expansion joints on long runs of gutter in order to avoid problems with expansion and contraction of the run.

- Make allowances for movement by properly aligning hangers and placing them at least 18 inches from corners adjacent to long runs.

- When the gutter ends are not confined to prevent excessive lateral shifting, secure the gutter to prevent creeping.

- This can be accomplished by drilling a nail hole as high as possible in the midpoint of the back wall of the gutter section.

- Then nail the gutter to the fascia, using shims if necessary, to maintain the proper alignment. Where the ends of the gutter are confined between walls or rake boards, maintain a clear space at each end to prevent water from standing between the gutter end and the wall covering.

- On buildings with inside or outside corners, always start a gutter system installation at a corner.

- Attach the gutter end securely to one corner before placing the section into the hangers.

- Once the section is clamped firmly into the hangers, attach the other gutter end.

- On a small hip-roof building, plan for the required expansion-contraction allowance using properly spaced hangers only.

- This is done by placing, on successive corners, one hanger 6 inches from a corner and the next hanger 18 inches from and around the corner.

- If this practice is followed on each corner, allowance is made for the typical thermal motion.

- As a rule of thumb, you will use rivets and polyvinyl chloride (PVC) cement to fasten sections together.

- Installation procedures for elbows and downspouts are similar to those used for metal sections.

- Always follow the manufacturer's assembly directions with care.

- Use only aluminum screws or nails.

- Other types of fasteners can corrode and stain the vinyl.

- Coat each screw or nail head with PVC cement to make it match the rest of the drainage system.

Replacing Gutters and Downspouts

When replacing gutters, remove the run all along one side of the house; otherwise you may distort the old gutters when you remove them. The old gutter run is very useful as a guide for fitting the replacement sections and joints.

- Once the old gutter is on the ground, line up the appropriate fixtures and the required number of new gutter lengths parallel to it.
- Mark the new pieces and cut them to the appropriate size.
- Use a carpenter's square to mark the cutting lines.
- Use a hacksaw or power saw with a metal-cutting blade for metal gutters. File the edges and apply a coat of rust inhibitor to the cut surfaces.
- Cut plastic gutters with a sharp, fine tooth handsaw.
- To prevent the gutter wall from flexing while it is being cut, slide a wooden filler block into the gutter and position it as close to the cutting point as possible.
- After cutting, file the rough edges.
- Before installing the new gutters, check the fascia board for rot.
- If it is defective, replace it. Be sure to paint both sides and the edges of the replacement board.
- Locate the downspout end of the gutter at the lowest point on the fascia board.
- To achieve this, snap a chalkline on the fascia board to show the desired slope for the gutter.

SOLVING PONDING PROBLEMS

- Ponding water on a roof is the number one enemy of the serviceable life of any low-slope roof system.
- Ponding water is the direct result of improperly located roof drains and drain supports placed directly on joists.
- After the building settles, drains rise well above the roof's drainline.
- In the 1950s, totally flat roofs were typically of a boilerplate design, usually with unsatisfactory results.
- Unless camber, or deflection, was built into the structural system to allow for anticipated dead and live loads, the deflection between bearing points resulted in low points on the roof.
- Typically, roof drains were located along beams or bearing walls, which end up being the high points in the roof system.

- This left little, if any, room for construction error.
- Everything, from the footings on up, had to be constructed to very close tolerances in order to end up with equal elevations across the building.
- The roofing industry gradually learned that ponding water acted to reduce the serviceable life of a built-up roof (BUR).
- Much more water entered the building if leaks did develop, and getting the roof dry enough to patch a leak was also a problem.
- Most major material manufacturers in the United States eventually came to recommend that slope be applied to all new roof designs using BUR membranes or manufactured single-ply membranes.
- There are many buildings, however, that still have no slope and excessive water ponding.
- The most opportune time to correct this condition is when reroofing becomes necessary.
- Ideally, eliminating all ponded water would be best for the long term performance of the roof system.
- There are times, however, when this is not practical because of the complexity of the situation or because of limitations imposed by the existing conditions.
- It is important to keep the design as simple as possible.
- A complex design might be the most effective, but costs rise proportionately, and the roofer might have difficulty executing the work.
- In these cases, the cost to eliminate the ponds exceeds the advantages of doing so.
- Many experts believe that a roof pond that dries within 48 hours after a rainfall has little effect on the long-term performance of a roof membrane.
- Once the existing conditions and the causes of the ponding problem have been determined, approaches to the reroofing design can be developed.
- Often, the addition of roof drains or scuppers is not feasible.
- Adding insulation of varying thicknesses, not necessarily tapered insulation, can be a viable solution in some cases.
- In response to the industry's movement toward sloped roof systems for complete drainage, manufacturers are marketing tapered insulation to give slope to relatively large sections of a roof area.

- Tapered insulation board is available in several different types, including polystyrene, perlite-aggregate, isocyanurate, cellular glass, and fiberglass.

- The slope available with each type of tapered insulation varies.

- The selected slope should be sufficient to achieve positive drainage, yet not so great that installing the insulation is difficult and costly. It may not be necessary to use tapered insulation to slope the entire roof area.

- There might be an existing deck slope, no matter how slight, that can be enhanced. However, it is best to bring in an architect to make decisions for the principal, removing design responsibility from the roofing professional's shoulders and placing it where it belongs.

- Let us look at a simple example of a 50-year-old building with an essentially flat concrete deck.

- The deck has gradually deformed and sagged between supports through the slow process referred to as creep.

- To compound the problem, new heating, ventilation, and air conditioning (HVAC) equipment has been added to the deck.

- The net result is a slight slope in the roof section from the supports to the center.

- The roof is drained at one roof edge through scuppers and over one of the bearing walls.

- The result is substantial ponding in the central portion of the roof.

- The reroofing design might include tapered insulation from the scuppers to the opposite roof edge to eliminate ponding.

- Since half of the existing roof already slopes toward the scuppers, however, why not take advantage of it?

- Placing flat insulation over this half of the roof gives a net positive slope toward the scuppers.

- Tapered insulation with enough slope to counteract the negative deck slope on the other half of the roof provides a positive slope across the entire roof area at the membrane level.

- This approach minimizes the insulation thickness so that roofers can better deal with height limitations at existing flashings, such as through wall flashings and at large mechanical units.

- Likewise, blocking requirements might be less intensive.

- The final product provides the desired drainage at less cost than a fully tapered system.

- Another typically inexpensive method to reduce ponded water is to increase the thickness of insulation in steps, or tiers, from the drainage points to the perimeter area.

- These tiers are typically made in increments with tapered edge strips that provide a smooth transition from one tier to the next.

- The same basic concept as in the first example is applicable here.

- Take advantage of the existing deck slope on half of the roof. In this case, the insulation is stepped down in four increments rather than having a continuous taper (Figure 12.11).

- The steps are located more closely in the area of greatest ponding, or negative slope, to minimize the existing highs and lows.

- The disadvantage to this approach is that it does not completely eliminate the ponding in all cases.

- It can reduce the depth of ponding to a tolerable level, however, by displacing the water in gradual increments.

- As with the tapered insulation example, height restrictions at the details are more workable with this method.

- Cost is the primary advantage to this approach.

- Depending on the type of roof system and the complexity of the insulation layout design, the installed insulation cost for the tiered system can be 10 to 40 per cent less than the tapered system.

- The cost savings can go a long way to offset any loss in service life of the roof system caused by minor ponding with the tiered system.

- Once the surface water gets to the drain, it must be able to flow off the roof.

- The extra materials used to flash in the drain or scupper often create a buildup around the drain, which results in ponding water.

- This effect can be minimized by placing the drain in the center of a depressed area that is enhanced by tapered insulation.

- The tapered area should be slightly larger than the largest flashing sheet.

- In climates where freezing temperatures often occur, ice damming at scupper outlets and ice damage to the downspout are often problems.

- The open-faced downspout was developed to minimize the potential for plugging while continuing to provide the wall protection and waterflow control afforded by a totally closed downspout.

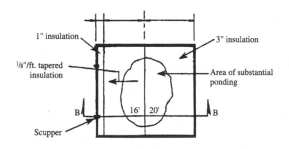

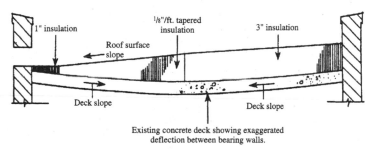

Existing concrete deck showing exaggerated
deflection between bearing walls.

FIGURE 12.11 Insulation stepped down in four increments.

- Using a dark color for the downspout helps solar heating keep the downspout as ice free as possible.
- Certain types of buildings, such as schools, can have other special considerations.
- It is amazing how a new downspout can provide children with access to the roof.
- A W-shaped, open-faced downspout securely anchored to the wall is a design that does not leave much for children to grab.
- Eliminate all projections or sharp sheet metal edges that can cause injury.
- The W-shaped design has an advantage over the more conventional open-faced downspout when factory color-coated sheet metal is used, as the entire exposed face shows the same run of color.
- The inside of the more conventional open-faced downspout shows the reverse side of the metal and produces a "racing stripe" on the building unless the inside is painted.

- This "childproof" downspout provides a reasonable solution to several common downspout problems.

- Many roof contractors have found the W-shaped downspout easy to fabricate and install, and they are satisfied that it provides essentially the same function as the more conventional open-faced downspout.

- Likewise, building owners are satisfied with the reduced potential for injury and traffic on the roofs.

- A better approach to edge drainage in northern climates is to install an inside/outside drain.

- This is usually more expensive than an exterior scupper and downspout, but its ability to remove snowmelt is much better, and the risk of damage from accumulated ice falling off the wall is eliminated. It is also relatively vandal proof.

- This type of roof drain system is simply a roof drain installed at the low point of the roof surface.

- The drain leader runs down the inside of the exterior wall and discharges through the wall 1 or 2 feet above grade (Figure 12.12).

- The interior location keeps the drain and leader open during cold weather.

- Ice buildup at the discharge is minimized because the drainage water has warmed slightly.

- Damage from any ice development is minimal because the ice forms at grade level and not at roof level.

- The best locations for roof drains depend on several conditions, including the highest water level, and good accessibility from the ceiling or along the predominantly low side of the roof.

- Since a roof membrane installation starts at the drain, it is ideal to line up new and old drains whenever possible.

- This way, the plies in the roof system do not buck the flow of water.

- From a cost standpoint, the best time to install a drain is while the old roof is being replaced.

- If the roof is in fairly good condition but suffers from large areas of standing water, definitely add new drains.

- Getting rid of ponding water reduces leaks, makes the roof last longer, and pays for itself in the long run.

- Select the drain that best fits the building conditions and the existing roofing system.

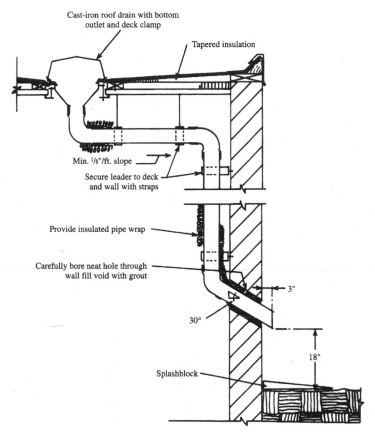

FIGURE 12.12 Inside-outside roof drain system.

- Drains can be made of cast iron, plastic, or copper.
- Consult local roofing supply houses for alternatives.
- To ensure that the proper components are obtained, use the checklist in Figure 12.13.
- On new, large construction jobs, the plumbing contractor may be responsible for the installation of the internal drainage system.
- In addition to providing and installing the roof drain, the plumbing contractor should ensure that the major components, such as a clamping device and grate, are made available to the roofer.

✓	Item
	Appropriately sized drain body
	Roof membrane clamping device
	Underdeck clamping assembly
	Expansion connection coupling
	Drain grate
	Roof-membrane flashing skirt

FIGURE 12.13 Gutter component checklist.

- The plumbing contractor's line of responsibility is normally completed once the drain is securely fastened to the roof substrate.

- Enter the roofing contractor, who must properly disassemble the parts, cut insulation, install flashing, and proceed with the installation of the roofing materials.

- The roofer must then fasten the clamping device securely over the in-place membrane.

- Future problems arise for the roofing contractor if the architect or specifier has not carefully selected the proper roof drain or the plumbing contractor has not installed it to specifications.

- This situation is slowly changing through grassroots efforts by some drain manufacturers and others in the roofing industry.

- Because the roofer is responsible for the drains in the end, it makes sense for him or her to be given the task in the contract and be paid for their installation.

- Manufacturers who now offer their own drains to accompany their roofing systems have recognized this and are the catalysts, at least in offering the roof drains to the roofing contractor.

- Perhaps in the future, roof drains will be entirely the responsibility of roofers.

RETROFITTING DRAIN SYSTEMS

- For retrofit projects, the pre-job walk through should include a look at the existing drains to determine whether a corrosion problem has occurred with the bolts.

- During the walkthrough, ask yourself some questions.
- Can you, in fact, count on the added time and labor involved with retapping, or should a complete drain replacement be addressed?
- Is the existing drain compatible with the specified roofing system?
- Address these questions beforehand so that you have more control over both the job and the associated job costs and potential profits.
- Discuss your concerns with the project manager or building owner before they become an issue.
- This enables you to better serve the client.

chapter

13

RESOURCES

ASPHALT ROOFING

Asphalt Roofing Shingles:
http://home improvement online.com/asphalt roofing shingles/

Asphalt Shingles: http://eetd.lbl.gov/coolroof/asshingl.htm

Asphalt Shingles and Roofing Materials from Thomas Regional:
http://www.materialstobuildit.com/shingles.html

Cascade Roofing & Waterproofing: http://www.bc biz.com/cascaderoofing/

Celotex Corporation: http://www.celotex.com

CertainTeed: http://www.certainteed.com/pro/roofing/ctroof/websites.html

Elk Premium Roofing: http://www.elkcorp.com/

Ondura Corporation: http://www.ondura.com/

San Tred Quality Flexible Coatings: http://www.sanitred.com

Sunshine Wholesale Roofing & Building Supplies, Inc:
http://www.sunshineroofing.com/

TAMKO Roofing Products, Inc.: http://www.tamko.com/

TRED Unique Liquid Rubber Roofing Products:
http://www.sanitred.com/Roofing.htm

Unique, High Roofing Products Company, Inc.: http://www.gsroof.com

ASSOCIATIONS AND AGENCIES

Adhesive and Sealant Council:
7979 Old Georgetown Road, Suite 500
Bethesda, Maryland 20814
Phone: (301) 986-9700
Fax: (301) 986-9795

Asphalt Institute:
Research Park Drive, P.O. Box 14052
Lexington, KY 40512-4052
Phone: (606) 288-4960
Fax: (606) 288-4999
http://www.asphaltinstitute.org/aboutai/about.htm

Asphalt Roofing Manufacturers Association:
4041 Powder Mill Road, Suite 404
Calverton, MD 20705
http://www.asphaltroofing.org/contact.html

Associated Roofing Contractors of the Bay Area Counties:
http://www.arcbac.org

ASTM:
100 Barr Harbor Drive
West Conshohocken, PA 19428-2959
Phone: (610) 832-9585
Fax: (610) 832-9555
http://www.astm.org/CONTACT/

The Cedar Shake and Shingle Bureau (CSSB):
P.O. Box 1178
Sumas, WA 98295
(604) 462-8961
http://www.cedarbureau.org/index.htm

Chicago Roofing Contractors Association: http://www.crca.org/index.html

Construction Estimating Institute: http://www.estimating.org/

Construction Technology Centre Atlantic:
http://ctca.unb.ca/CTCA/document/estimate.html

Environmental Protection Agency:
http://www.epa.gov/epahome/comments.html

About the EPA Public AccesServer:
http://www.epa.gov/epahome/about.html

EPA, Clean Air, and Small Business:
http://sbinformation.about.com/library/weekly/aa072897.htm?iam=m

Search the U.S.: Search the EPA Internet site: http://search.epa.gov/

EPA's Office of Solid Waste: http://www.epa.gov/epaoswer/osw/

Midwest Roofing Contractors Association: http://www.mrca.org/

North/East Roofing Contractors Assn., Inc.
http://ww1.thebluebook.com/ne/htm/0011672100000.shtml

Midwest Roofing Contractors Association: http://www.mrca.org/

The National Roofing Contractors Association: http://www.roofonline.org/
Occupational Safety and Health Administration (OSHA):
http://www.osha slc.gov/SLTC/asbestos/index.html

Roofing and Sheet Metal Contractors Association of Georgia:
http://www.rsmca.org

Roofing Contractors Association of Washington: http://www.rcaw.com/

Roofing Intelligence Member Associations:
http://www.roofingintelligence.com/assoc.htm

Small Business Association (SBA): http://sba.governmentloans.com/

SBA Ombudsman: http://www.sba.gov/regfair/

SBA Introduction: http://www.sbaonline.sba.gov/intro.html

Stone Roofing Association: http://www.brookes.ac.uk/geology/stoneroof/

Western States Roofing Contractors Association: http://www.wsrca.com/

BUILT-UP ROOFING

ADURON: http://www.aduron.com/

Atlas Roofing Corporation: http://www.atlasroofing.com/

BAJA Built-Up Roofing Systems: http://www.bajaroofing.com/builtup.html

Built-Up Roofing Specifications:
http://www.comforttherm.com/roofing/builtuproof/specifications

CBD 24. Built Up Roofing: http://www.nrc.ca/irc/cbd/cbd024e.html

EJ Roberts Roofing Limited: http://www.e j roberts roofing.co.uk/felt.htm

JM Roofing Systems: http://www.comforttherm.com/roofing/builtuproof

Physical Maintenance: http://facilitiesnet.com/NS/NS3mg6i.html

River Roofing: http://www.riverroofing.com/hottar.html

Roofing Products: http://www.jpselastomerics.com/roofing/index2.html

Sun Life Systems: http://www.sunlifesystems.com/

Tamko: http://www.tamko.com/

CLAY TILE ROOFING

AmeriClay Roof Tile: http://www.americlay.com/

Cement roofing tiles:
http://www.nrc.ca/irc/thesaurus/cement_roofing_tiles.html

eBuild: http://www.ebuild.co.uk/products/roofing.htm

International Roofing Products, Inc:
http://www.hybgroup.com/irpweb/htm/

Eagle Roofing Products: http://www.eagleroofing.com/

Elagante' Roofing Tiles: http://www.elagante.com/

GCI Clay Roof Tiles: http://www.goldenclay.com/aboutus.html

Golden Clay Industries: http://www.goldenclay.com/

Interclay Corp.: http://www.altusa.com/

James Hardie: http://www.JamesHardie.com/

Kia Lim: http://www.roof tiles.com/

Ludowici Roof Tiles: http://www.ludowici.com/

Marley Roofing Products: http://www.marleyroofing.co.uk/

MCA Clay Roofing Tile: http://www.mca tile.com/

Mineral Fiber Cement Roof Tile Companies:
http://www.building.com/communities/texis/db/go/+BwwrmwAeufzw/com

Mission San Luis Obispo De Tolosa:
http://www.cuca.k12.ca.us/lessons/missions/slo/SanLuisObispoDeTol

Northern Roof Tile Sales Company, Inc.: http://www.northernrooftiles.com/

Polet IGK: http://www.polet.co.yu

Redland Tile & Brick: http://www.redland tile brick.co.uk

Renaissance Roofing, Inc.: http://www.claytileroof.com/

Roofing: http://www.greenbuilder.com/sourcebook/Roofing.html

Sandtoft Roof Tiles: http://www.sandtoft.co.uk/index.html

Tudor Roof Tiles Co. Limited: http://www.tudorrooftiles.co.uk/

U.S. Tile Company: http://www.ustile.com

DEWATERING EQUIPMENT

Piranha Pumps: http://www.piranhapumps.com/

Griffin Dewatering Corp. Texas:
http://www.thebluebook.com/fs/htm/0037207011000.shtml

National Manufacturers of Pumps:
http://www.constructionnet.net/manu_search/results/55221.shtml

Wanner Hydra Cell Pumps: http://www.texaspump.com/

ENVIRONMENTALLY RESPONSIBLE ROOF SUBJECTS

Big Green Machine: http://www.empnet.com/BigGreen

Renewable Energy Policy Project:
http://solstice.crest.org/efficiency/strawbale list archive/9702/

Winpower Generators: http://www.winpowerinc.com/

GENERATORS

Americas Generators: http://www.americasgenerators.com

Best Power Generators: http://www.best power generators.com/

Bowers Power Systems: http://www.bowerspower.com/

Circle H Generators, Inc.: http://www.circlehgen.com/

Fischer Panda: http://www.fischerpanda.com/

Generac Power Systems: http://www.generac.com/

Generators: http://www.cairns.net.au/~sharefin/Markets/Alt2.htm

Generators: http://generatorsunlimited.com/

House of Generators: http://www.powerpony.com/

Master Generators: http://www.mastergenerators.com/

Portable Generators from USA Light & Electric:
http://www.yamahagenerators.com/

Power Generators: http://globalgenerator.com/index.html

PTO Generators: http://www.ssbtractor.com/PTO_generators.html

Weber Generators: http://www.webergenerator.com/web13/

Winpower Generators: http://www.winpowerinc.com/

Young Generators, Inc.: http://www.younggen.com

GUTTERS AND ACCESSORIES

Ryan Seamless Gutters: http://www.ryangutter.com/

Weather Guard Building Products: http://www.wgbp.com/

INSULATION

Polycoat Systems, Inc.: http://www.polycoat.com/framset2.html

Sprayed Polyurethane Foam Roofs: http://www.spf roofs.com/

Team Industries, Inc.: http://www.teamindustries.com

MAGAZINES

Construction Magazines Online:
http://www.geocities.com/Paris/Louvre/1776/magazines_construction

Iroofer.com: http://www.iroofer.com/tile.htm

Professional Roofing Magazine:
http://www.professionalroofing.net/past/may99/

Roofing Contractor Online: http://www.roofingcontractor.com/

Roofing Magazine: http://www.build.co.uk/roofing.htm

Steel Roofing: http://www.steelroofing.com/where.htm

MARKETING

AEC Resource Guide: http://www.nwbuildnet.com/nwbn/resourceguide.html

Architecture Shopper:
http://architecture.about.com/library/blq%26a.htm?iam=ma

B & M Steep Roofing Profile:
http://www.building.org/texis/db/bix_search/+fnWePY85mrmwwweR2zw/p

Build.com Home Page: http://www.build.com/

BuildingOnline: http://www.buildingonline.com/blsearch.shtml

Directory of Building and Roofing Business Opportunities:
http://www.infoarea.com/vitamins/small business information/58.ht

Holland Roofing, Inc.:
http://www.building.org/texis/db/bix/+0wwrmwxerd1wrmwxCeuxww/prof

Remodel Online Home:
http://www.remodelonline.com/directories/construction/roofing/gen

RoofHelp: http://www.roofhelp.com

roofing.com: http://www.roofing.com/

Roofing Consultants Online: http://www.iroofer.com/

Roofing Webbing Homepage: http://www.latch.com/webring.html

RSI Roofing, Siding, Insulation:
http://www.advanstar.com/markets/roofing.cfm

S & K Roofing Profile:
http://www.building.org/texis/db/bix/+0wwrmwxerA1wrmwxexhMw/prof

METAL ROOFING

Advanced Metal Roof Systems: http://www.amrs.w1.com/

AEP SPAN Metal Roofing Solutions: http://www.aep span.com/

Copper Page: http://www.copper.org/

Custom Bilt Metals: http://www.custombiltmetals.com/

Lianro Metal Roofs, Inc.: http://www.lianro.com/

Metal Roofing Tools and Equipment:
http://www.custombiltmetals.com/tools.htm

Pioneer Roofing Systems: http://www.pioneerroofing.com/

RECYCLING

Authentic Roof TM: http://www.authentic roof.com/

Asphalt and Tar Recycling:
http://www.recycle.net/recycle/spec/gr080215.html

Big Green Machine: http://www.empnet.com/BigGreen

Specification Chemicals, Inc: http://www.spec chem.com/rrr/

RENTALS AND LEASING

Leasing Options: http://www.4equipmentleasing.com/? %3B019039a

Roofing Equipment: http://www.wirtzrentals.com/roofing.htm

Roofing Equipment: http://www.fastrentals.com/roofing.html

ROOFING FABRIC

Bondcote: http://www.bondcote.com/

ROOFING SUPPLIES AND MISCELLANEOUS

Heely Brown Roofing Equipment & Supplies:
http://heely brown.com/hbroof1.html

Reeves Roofing Equipment: http://www.reevesequipment.com/main.htm

Robseal Roofing Case Study: http://www.robseal.com/case1.htm

Roof Center: http://www.roofcenter.com

Roofers Mart: http://www.roofersmart.com/

Roofing.com: http://www.roofing.com/

Roofing Material Housemart.com: http://www.roofingmaterial.webscour.net/

Roofing Materials and Hot Rubber Roofing Materials:
http://kettles.web scape.com/roofing.html

Roofing Siding & Insulation Materials Wholesale:
http://www.okdirect.com/Biz/5/5033.html

Roofing Wholesale Co., Inc.: http://www.biw.co.uk/BIW/register/2510.htm

Shop CRS: http://www.shopcrs.com

Stoneway Roofing Supply: http://www.stonewayroofing.com

Structural Materials: http://www.roofs.net/smc/index.htm

ROOFING SYSTEMS

Bondcote: http://www.bondcote.com/

GenFlex Roofing Systems: http://www.genflex.com/

Interlocking Roofing Tiles: http://www.goldenclay.com/rooftiles.html

Johns Manville Commercial & Industrial Roofing Systems:
http://www.jm.com/roofing/index.html

Pioneer Roofing Systems: http://www.pioneerroofing.com/

Polycoat Systems, Inc.: http://www.polycoat.com/

SAFETY

Ace Fire Protection: http://www.acefireprotection.com

Able First Aid: http://www.ableaid.com

Adventure Medical Kits:
http://www.adventuremedicalkits.com/home/home.htm

Ekman Safety: http://www.ekmansafety.com/index.html

Equipment Safety: http://www.4buildingsupplies.com/?%3B019039a

Fire Safety and Fire Extinguishers:
http://www.chem.uky.edu/resources/firefighting.html

First Aid Direct: http://www.firstaid direct.co.uk/

First Aid Kits: http://www.healthproductswh.com/hpw/firstaidkits.html

First Aid Kits: Johnson & Johnson First Aid Kits
http://www.easyshopn.com/first_aid_kits.htm

First Aid Kits: http://www.firstaidkits.lookandgetit.com/

GovSearch: http://www.nwbuildnet.com/nwbn/govbot.html

Halon 1211 Fire Extinguishers:
http://www.reliablefire.com/portablesfolder/halon_1211_i.html

Korkers: http://www.dumarbusiness.com/korkers.htm

North East Fire and Safety Equipment Co.:
http://www.northeastfireinc.com

OSHA Safety Training: http://www.burtonandassociates.net

Portable Fire Extinguishers: http://www.vbg.org/FIRE/Fire ext.htm

Portable Generators:
http://usgovinfo.about.com/library/news/aa122899a.htm?iam=ma&term

Pdxfire Website by TAS Mania Graphics: http://www.pdxfire.com/

Pyrotec Fire Protection: http://www.city2000.com/md/pyrotec.html

Quips Safety: http://198.252.9.95/oshaweb/IQ_data/IN3D4B~4.HTM

Safety Tips: http://www.vbg.org/FIRE/fire ext.htm

Wheeled Fire Extinguishers:
http://www.reliablefire.com/portablesfolder/wheeledunits.html

SLATE

American Slate Company: http://www.americanslate.com/

Aspigal Roofing Slate: http://www.aspigal.com/ie/index.asp

Black Diamond Slate Company: http://blackdiamondslate.com

Buckingham Virginia Slate Roofing: http://www.bvslate.com/roof.htm

Echeguren Slate, Inc.: http://www.echeguren.com

Evergreen Slate Company: http://www.evergreenslate.com

Greenstone Slate: http://www.greenstoneslate.com/

Historic Slate Roofs: http://www.neslate.com/Levine.html

International Roofing Products, Inc.: http://www.hybgroup.com/irpweb/htm/

Masterpiece Tile Company, Inc: http://www.masterpiecetile.com/

Renaissance Roofing, Inc.: http://www.claytileroof.com/

Roof Tile and Slate Company: http://www.rtsc.com

Schumacher Slate Works: http://www.slateroof.com/

Slatecraft: http://www.slatecraft.com/

Slate import and export from David Wallace Slate International Ltd.:
http://www.slate.uk.com/

Slate Layout: http://www.slatelayout.com/

Slate Roofing: http://llechwedd.co.uk/nfslate.htm

Slate Roof Bible: http://www.jenkinspublishing.com/slate.html

Stoneside Slate Roofing: http://www.stoneside.com/samp11.htm

The New England Slate Company: http://www.neslate.com/

The Slate Book: How to Design, Specify, Install, and Repair a Slate Roof:
http://www.oldworlddistributors.com/pix_slatebook.html

Universal Slate International Inc.: http://www.universalslate.com

Vermont Slate Sales, Inc: http://www.sover.net/~slate/index.html

Vermont Slate Roofs, Weathervane & Finials:
http://www.slateandcopper.com/

Virginia Slate: http://www.virginiaslate.com

SLATE TOOLS

Hytile: http://www.hytile.com

Jenkins Slate Store: http://www.jenkinsslate.com/store_tools.html

Roofing Slate Tools Machinery and Materials:
http://www.oldworlddistributors.com/toolsmaterialsindex.html

The New England Slate Company Tools & Hardware:
http://www.neslate.com/Tools.html

The Roof Tile and Slate Company: http://www.rtsc.com/tools_nails.html

SNOW GUARDS

Real Tool: http://bergerbros.com/real tool/Materials.html

Removing Snow and Ice from Roofs:
http://www.buffalo.bbb.org/alerts/snowremoval.html

SOFTWARE

AccountPro 98: http://www.accountpro.com/mod178.htm

AppliCad Australia: http://www.applicad.com.au/breaking.htm

ARCAT Section 07550: http://www.arcat.com/divs/sec/sec07550.cfm

ASR Home Page: http://www.asrsoft.com/

Baarns Publishing: http://archive.baarns.com/excel/products/costest.asp

Bruco Enterprises, Inc.: http://www.flash.net/~bruco

Xpertsite.com: http://www.xpertsite.com/ShowCategory.asp?category_id=483

Data Trak Maintenance and Roofing Software:
http://www.maintenance software.com

Material Estimating Spread Sheets:
http://www.geocities.com/SiliconValley/Heights/8519/download.html

Estimating Software for Restoration and Cleaning:
http://www.qirra.com/Simplicity.ht

Excel Spreadsheets in Estimating:
http://www.baanfans.com/baanfans/forum/project/messages/19.html

FastEST, Inc: http://www.fastest inc.com/excel.htm

FAST Planner for IT: http://www.rms.net/fast_planner_news.htm

Roofing Estimating Software:
http://www.jlconline.com/forums/computers/messages/513.html

Links to all construction cost estimating software downloads on the Web:
http://bidshop.org/

Microsoft Word contractor and distributor proposal and quotation software:
http://amplifyllc.com/onequot/contractor/index.html

Roof Wizard, Roof Magician and Cadair estimating software:
http://www.applicad.com.au/products.htm

Tempest Company Estimating Capabilities:
http://www.tempestcompany.com/est1.htm

Roofing Contractors Software by FunnelVision:
http://www.funnelvision.com/WEB_297.HTM

Roofing Software: http://www.tsnorthland.com/

Roofing Materials Suppliers, Building Products:
http://www.affordablewebdesign.com/roofing.htm

Roofing System: http://www.impomag.com/B_comp/TEMPLATE.HTM

Swiftest: http://www.swiftest.com/

UDA Spreadsheet Templates for Estimating Construction Costs:
http://www.uniteddesign.com/excel_spread.html

TOOLS, HAND

1999 Tool & Equipment Guide:
http://www.roofingcontractor.com/tool equipment_guide.htm

E.L. Hilts: http://www.elhilts.com/

AJC Tools and Equipment: http://www.ajctools.com

H & H Tool Co.: http://www.hamlettools.com

Leister Roofing: http://www.hotairtools.com/Roofing.htm

Metal Roofing Tools and Equipment:
http://www.custombiltmetals.com/tools.htm

Proshows: http://www.proshows.com/attendee/exhp&s.htm

Rocket Equipment: http://www.rocketequipment.com/

Roofing Tools RM Virtual Store: http://roofersmart.com/rmstore/ajc.htm

Roofstripper Roofing Tool:
http://www.sowashco.com/roofstripper/index.shtml

Roofing Tools and Roof Products by THOR Roofing Tools:
http://www.thorsystems.com/

Roofing with Senco Fasteners and Tools:
http://www.senco.com/tech_tips/nailsvsstaplesroofing.html

Rusko: http://www.rusko.com/

Stoney Cedar Roofing Tools: http://www.renweb.net/stoneycedar/tools01.htm

Tent, Tarp, and Roofing Equipment: http://www.wegener na.com/tent.htm

Werner Ladder Co.: http://www.WernerLadder.com

TOOLS, HEAVY ROOFING

Acro Building Systems, Inc.: http://www.acrobuildingsystems.com/

Cleasby Roofing Products: http://www.cleasby.com/

Commercial Roofing Equipment:
http://garlockequipc.hlkj.com/commercial_roofing_equipmentlbd/

Commercial Roofing Specialties, Inc.: http://www.crsroofingsupply.com/

Garlock Equipment Company: http://www.garlockequip.com/

Hoisting and On Deck Roofing Equipment:
http://heely brown.com/hoist.html

New Tech Machinery Corp.:
http://search.commerceinc.com/profile/id3108027136.html

Panther Products Roofing Equipment Online:
http://www.toto.net/panther/nonframe.html

Phoenix Sales, Inc.: http://www.phoenixsalesinc.com/about.htm

Reeves Roofing Equipment Inc.:
http://www.reevesequipment.com/distributors.htm

Roofing Equipment:
http://www.nrc.ca/irc/thesaurus/roofing_equipment.html

Roofing Equipment: http://garlockequip.hlkj.com/roofing_equipmentlba/

Roofing Equipment: http://www.dsidiamond.com/roofing.htm

Roofing Equipment: http://columbineint.com/prod_4.htm

Roofing Equipment: http://www.tiedown.com/roofing.htm

Roofing Equipment and Supplies:
http://www.chicagoconstruction.com/search/r htms/roofsup.htm

Roofing Machinery for Residential and Commercial Applications:
http://www.rollformers.com/pages/frameset.htm

Roofstripper: http://www.sowashco.com/roofstripper/index.shtml

WOOD SHAKES AND SHINGLES

3M Algae Block Copper Roofing Granule System:
http://www.3M.com/algaeblock/intro.html

Anglo American Cedar Products Ltd.: http://www.angloamerican.com

BCF Shake Mill Ltd.: http://www.bcfshake.com/shakes.html

D & R Cedar Products: http://dandrcedarproducts.uswestdex.com/

Fraser Cedar Products: http://www.frasercedarproducts.com/

Granville Manufacturing Company, Inc.: http://www.woodsiding.com/

Improve Net Types of Roofing Shingles:
http://www.improvenet.com/ref/rffs.html?RU=rfea.html&PID=19&U=htt

High Desert Trading: http://www.highdeserttrading.com/shakes.html

John Brash: http://www.johnbrash.co.uk/enquiry.html

LifePine Wood Shakes and Shingles: http://www.olesmoky.com/p_lifepn.ht

Meeker Cedar Products Ltd.: http://www.meekercedar.com/

Metal Shake Roofing: http://www.ahiroofing.com/shakes.htm

Metro Roofing Supplies:
http://www.metroroofingsupplies.com/wood_slate.htm

Monier Lifetile: http://www.monierlifetile.com/

Penofin: http://www.PENOFIN.com

ProTex Shake: http://www.bobsroofing.com/specialtyshingles.htm

Rejo Cedar: http://www.nwshakenshingle.com/

Stave Lake Cedar: http://www.stavelake.com/

Steadfast Cedar Shakes and Shingles: http://www.steadfast.com/

Teal Cedar Products, Ltd.: http://www.TealCedar.com/

Watkins Sawmills Limited: http://www.watkinsawmills.com

Wood Shingles Directories:
http://www.building.org/texis/db/bix/+jwwrmwAeuASwrmwxe2hww/commu

INDEX

A

Accounting, 1, 2

Aggregate, 137, 157, 178, 202, 209

Air-supported fabric roofs, 114-115

Algae discoloration in asphalt shingles, 276-278

Aluminum roofing materials (*see Metal roofing*), 405, 410, 411-413, 418, 435, 440, 441

American Society for Testing and Materials (ASTM), 2

Architect's responsibilities on roofing jobs, 18

Area calculations, roofs, 8

Asphalt, 117-158, 211
  heating temperatures, 138

Asphalt roll roofing, 211, 212, 215

Asphalt shingles over old roll roofing, 248
  double-coverage roll roofing, 269
  inspection, 275
  open valleys and flashing, 261
  parallel-to-eaves or rake, 261

Asphalt roofing
  algae discoloration, 278
  bridging techniques, 250-254
  chimney flashing, 256
  counterflashing in brick, 240-241
  cricket on roof, 239, 240, 250
  dormers, 228, 235, 256

drip edges, 244-248

eave flashing, 232, 242, 244, 247

first and successive course application, 242, 254-255, 262

flashing, 215, 217, 228-247, 274

giant shingles, 258

hexagonal shingles, 258

ice dams, 242

inspection, 275

interlocking shingles, 213-214, 259

low-slope roofs, 211, 246-247, 253, 270

mineral granule coatings, 212-213

nailing recommendations, 216

nesting techniques, 250-254

patterns for laying three-tab strip shingles, 214, 221, 224

racking or straight-up application method, 223, 277

random spacing pattern, 227-228

reroofing operations, 249

reroofing over old asphalt roll roofing, 252

reroofing over old asphalt shingles, 251

reroofing over old BUR roofing, 252

reroofing over old slate roofing, 253

reroofing with shingle strips, 248

reroofing over wood shingles, 251

ribbon courses, 227

roll roofing, 261

NOTES

NOTES